Thomas Bevill Peacock

On malformations of the human heart, etc.

With original cases and illustrations. Second Edition

Thomas Bevill Peacock

On malformations of the human heart, etc.
With original cases and illustrations. Second Edition

ISBN/EAN: 9783744736817

Printed in Europe, USA, Canada, Australia, Japan

Cover: Foto ©berggeist007 / pixelio.de

More available books at **www.hansebooks.com**

ON

MALFORMATIONS

OF THE

HUMAN HEART.

ON

MALFORMATIONS

OF THE

HUMAN HEART.

ETC.

WITH ORIGINAL CASES AND ILLUSTRATIONS.

BY

THOMAS B. PEACOCK, M.D.,

FELLOW OF THE ROYAL COLLEGE OF PHYSICIANS;
PHYSICIAN TO ST. THOMAS'S HOSPITAL; AND SENIOR PHYSICIAN TO THE HOSPITAL FOR
DISEASES OF THE CHEST VICTORIA PARK;
PRESIDENT OF THE PATHOLOGICAL SOCIETY OF LONDON.

SECOND EDITION.

LONDON:
JOHN CHURCHILL AND SONS, NEW BURLINGTON STREET.
MDCCCLXVI.

LONDON :
SAVILL AND EDWARDS, PRINTERS, CHANDOS STREET,
COVENT GARDEN.

PREFACE.

October, 1858.

SEVERAL examples of malformation of the heart having fallen under my notice, I have for some years paid special attention to this branch of pathology ; and in 1854, shortly after an interesting case of malformation had occurred at St. Thomas's Hospital, I delivered a series of lectures on the subject to the students. These lectures were subsequently published in the " Medical Times and Gazette ;" and, together with several cases which have been at various times contributed to different medical societies and journals, are now reprinted, after the whole has been carefully revised and considerably extended.

There are few subjects which have attracted more attention in the profession than the irregularities in the development of the heart and large vessels. Longer or shorter notices of the chief varieties of malformation are contained in different works on Morbid Anatomy, and in systematic treatises on cardiac affections by Burns, Corvisart, Bertin, Laennec, Bouillaud, and Hope. They have been made the subject of a special essay by Dr. Farre,[1] and of

---

[1] On Malformations of the Human Heart. London, 1814.

b

shorter memoirs by Dr. Paget,[1] Dr. Williams,[2] Dr. Todd,[3] Dr. Joy,[4] and Dr. Craigie ;[5] by Haase,[6] Meckel,[7] and Hein ;[8] and by Louis[9] and Gintrac.[10] More recently the subject has been discussed in graduation theses by MM. Deguise[11] and Pize,[12] and in a series of valuable and interesting papers by Dr. Chevers.[13]

It might thus appear that there was scarcely scope for a new work on the subject. Several of the memoirs however which I have named, were published at a period when the defects in the conformation of the heart were less studied than they have recently been. Others are limited to one department of the subject; and the treatise of Dr. Farre is

---

[1] On the Congenital Malformations of the Human Heart. Edin. Med. and Surg. Journ., vol. xxxvi. 1831, p. 263.

[2] Cyclopædia of Practical Medicine, vol. iii. 1834, p. 65.

[3] Cyclopædia of Anatomy and Physiology, vol. ii. 1839, p. 630.

[4] Library of Medicine, vol. iii. 1840, p. 381.

[5] Edin. Med. and Surg. Journ., vol. lx. 1843. Case of Cyanosis, or Blue Disease, p. 265.

[6] Dissertatio Inauguralis Medica de Morbo Cœruleo. Lipsiæ, 1813.

[7] De cordis conditionibus abnormibus Dissertatio Inauguralis. Halæ, 1802. Beitrag zur Geschichte der Bildungsfehler des Herzens welche die Bildung des rothen Blutes hindern. Deutsches Archiv f. d. Physiologie, Halle und Berlin, 1815, p. 221.

[8] De istis cordis deformationibus quæ sanguinem venosum cum arterioso misceri permittunt. Gœttingæ, 1816.

[9] Archives Générales de Médecine, 2mo série, t. iii. 1823 ; Mémoires ou Recherches Anatomico-Pathologiques. Paris, 1826, p. 300.

[10] Observations et Recherches sur la Cyanose, ou Maladie Bleue. Paris, 1824.

[11] De la Cyanose Cardiaque, etc. Thèse de Paris, 1843.

[12] Considérations sur les Anomalies Cardiaques et Vasculaires qui peuvent Causer la Cyanose. Thèse de Paris, 1854.

[13] Collection of Facts illustrative of Morbid Conditions of the Pulmonary Artery. London, 1851, originally published in London Medical Gazette, 1845 to 1851. The causation of Cyanosis is also discussed by Ferrus in the art. Cyanose, Dict. de Méd., 2mo ed., t. ix. 1835, p. 527 ; by Stillé on Cyanosis or Morbus Cæruleus, in Am. Jour. of Med. Sc., N.S., vol. viii., Phil. 1844, p. 25; and by Copland, in the Dictionary of Practical Medicine, art. Blue Disease, vol. i. p. 199.

professedly incomplete. I have, therefore, thought that a work containing the more recent information would not be without interest and value to the profession.

The subjects treated of in the Essay embrace :—

    I. Congenital Misplacements of the Heart.
    II. Deficiency of the Pericardium.
    III. Malformations of the Heart.
    IV. Irregularities of the Primary Vessels.
    V. Mode of formation ; Symptoms and Effects ; Diagnosis and Medical Management of cases of Malformation.

In the Essay it has been my aim to present the subject in a practical point of view, and I have not, therefore, alluded to those forms of defect in the development of the heart which are incompatible with the existence of extra-uterine life, or which have only been met with in the lower animals. The medical periodicals of Germany, France, England, and the United States, abound in cases of malformation. I have, however, not thought it desirable to quote many of these at length, but have preferred to allude concisely to those which formed the earliest published examples of each form of defect, or which presented some rare and remarkable deviation, and have generally contented myself with referring to the periodicals and other publications in which the more recent or ordinary cases are reported. A brief inspection of the numerous references cited, will show that to have quoted more extensively would have greatly swelled the bulk of the volume. Those who desire to investigate the subject further, I may refer to the original

sources of information here indicated; or to the essay of
M. Gintrac, the theses of M. Deguise and M. Pize, and the
papers of Dr. Chevers; in one or other of which, but particularly the latter, a large proportion of the cases, only
briefly noticed or alluded to in this work, will be found
quoted at length.

October, 1866.

THE former edition of this work was a small one, and was
rapidly disposed of, so that it became out of print in a
short time. The present edition has been carefully revised,
several cases and illustrations have been added, and the
references have been considerably increased. In its preparation I have availed myself of various treatises and
memoirs which have been recently published or which I
had not previously seen. I may particularly mention the
works of Friedberg,[1] Tiedemann,[2] and Förster,[3] and the
papers of Professors Mayer[4] and Kussmaul,[5] and of Dr. Carl
Heine.[6] It should also be mentioned that, in addition to

---

[1] Die angebornen Krankheiten des Herzens und der grossen Gefässe des Menschen.
Leipzig, 1844.

[2] Von der Verengung und Schliessung der Pulsadern in Krankheiten. Heidelberg and Leipzig, 1843.

[3] Die Missbildungen des Menschen. Jena, 1861.

[4] Virchow's Archiv für path. Anat. und Phys., etc., xii[ter] Band, 1857, pp.
364, 495.

[5] Henle and Pfeuffer's Zeitschrift für Rationelle Medicin, xxvi[ter] Band, 1865,
p. 99. Reprinted at Freiberg, I.B., 1865.

[6] Angeborene Atresie des Ostium arteriosum dextrum. Tübingen, 1861.

the references before given, longer or shorter notices of the various malformations of the heart and vessels have also appeared in the treatises of Walshe, Fuller, and Markham.

Notwithstanding that some very remarkable deviations from the natural conformation of the heart were placed on record at an early period, it is only recently that attempts have been made to reduce the different forms of irregular development to any scientific arrangement, or to explain their nature and mode of production. The earlier writers on cardiac affections, Corvisart, Laennec, Burns, Bertin, &c., contented themselves with simply alluding to the different forms of defect with which they were acquainted; and even so recent a writer as Dr. Hope regarded them as "so irregular in their combinations as scarcely to admit of being classified on general principles." Dr. Farre adopted an arrangement into, 1st, Malformations of the heart or arteries occasioning the mingling of black with red blood; and, 2ndly, those causing impediment to the circulation of the blood. The former class he further subdivided according to the degree of imperfection in the development of the heart. Dr. Paget and Andral described the various malformations as consisting in defective, excessive, or perverted development; and most subsequent writers have followed a similar arrangement. M. Bouillaud has treated of the various congenital anomalies of position and development of the heart and its vessels under the heads of—1st. Dexiocardia, or transposition. 2ndly. Communications between the right and left cavities of the heart from non-obliteration of the foramen ovale or ductus arteriosus, or the existence of a perforation in the septum of the auricles or ventricles. 3rdly. Anomalies of number, defective or excessive development of the heart or its several parts ; and, *4thly*. Anomalies of reciprocal connexion and insertion of the heart and its vessels. M. Berard[1] classifies the cardiac malformations into—1st. Anomalies of number; 2nd. Of direction and relation; 3rd. Those which do not cause the admixture of arterial and venous blood ; and, 4th. Those which are attended by such intermixture.

Friedberg, whose treatise, though published in 1844, I had not seen when the first edition of this work appeared, arranged the different forms of congenital malformation into three classes, founded upon the periods at which the process of development of the ovum is arrested or perverted. 1st. Those in which the formation of the partition-wall in the heart and the common arterial trunk has not taken place ; or anomalies belonging to the first period of fœtal development. 2nd. Those where the partition-wall is imperfectly formed or irregularly arranged, causing abnormal communications below the several cavities of the heart, or faulty origin of the vessels : these cases originating in the second period of development ; and, 3rdly. Those where the orifices and their valves are defectively constructed, or the connexion with the aorta continues ; defects which have their explanation in derangements of the formative processes of the third period.

Dr. Carl Heine, a recent writer on the subject, proposes a classification, founded on the systems of Förster and

---

[1] Art. Anomalies du Cœur, Dict. de Méd., 1834.

Bischoff, into—1st. Malformations which deviate quantita-
tively from the type of their kind; and, 2ndly. Those which
differ qualitatively. The former he subdivides into anomalies consisting in defective and excessive development as to
number and size; the latter he further distributes into
deviations of form and position, and irregularities of the
large vessels. The defective developments he classes under
the heads of—*a*. entire defects; *b*. imperfect formations;
c. abnormal smallness; *d*. obliteration of the passages or
vessels, and *e*. fissure formations.

In the present edition of this work, I have adhered to
the arrangement of the misplacements of the heart proposed
by M. Breschet; treating, first, of the cases in which the
organ does not occupy its proper position within the thorax;
and, secondly, of those in which, from defect in some portion of the parietes, the heart is situated wholly or in part
external to the thoracic cavity. I have, however, followed
Dr. Alvarenga[1] in the employment of terms, in some cases
different from those of M. Breschet.

Of the malformations of the heart itself, I adopted,
in the former edition, an arrangement founded partly on
the period at which the development of the organs becomes
arrested or perverted, and partly on the degree of impediment to the circulation which such deviation occasions, and
the consequent interference with the functions of the heart
after birth. To this arrangement, though with some modification as to details, I still adhere.

---

[1] See paper read before the Academy of Sciences in Lisbon, 1866.

The cardiac anomalies are, therefore, classed into—

1st. Those dependent on arrest of development at an early period of fœtal life, or probably from about the fourth to the sixth week, so that the organ retains its most rudimentary forms;—the auricular and ventricular cavities being still single or presenting very slight indications of division, and the primitive arterial trunk being retained, or the aorta and pulmonary artery very imperfectly evolved.

2ndly. Those in which the arrest of development occurs at a more advanced period of fœtal existence, or probably from about the sixth to the twelfth week, when the auricular and ventricular partitions have already been partly formed, and the aorta and pulmonary artery more or less completely developed. Such are the cases in which, with imperfect separation of the ventricles, the arterial or auriculo-ventricular apertures are constricted or obliterated, and the origins of the primary vessels, and especially of the aorta, are misplaced.

3rdly. Cases in which the development of the heart has progressed regularly till the later periods of fœtal life, so that the auricular and ventricular septa are completely formed, and the primary vessels possess their natural connexions; but in which, from the occurrence of disease, the organ is either prevented undergoing the changes which should ensue after birth, or there are slighter morbid conditions which may become the source of serious obstruction at more advanced periods of life. Such are the premature closure of the fœtal passages, and the occurrence of irregularities in the number and form of the valves.

The irregularities of the primary vessels may be similarly classed into—

1st. Those taking place at the earlier periods of fœtal life, and consisting in the defective evolution of the aorta and pulmonary artery from the primitive vessel and branchial arches.

2ndly. Those in which the development of the aorta and pulmonary artery is less deranged, but in which there are defects which may give rise to serious results in after-life.

The former class includes the cases in which the origins of the pulmonary artery and aorta are transposed, or the descending aorta is wholly or in part derived from the pulmonary artery. Of the latter class the most striking examples are the cases in which the aorta beyond the origin of the left subclavian artery is more or less constricted, and so occasions disease which may ultimately lead to very marked contraction or even entire obliteration of the canal.

From the arrangement of the various congenital defects into misplacements, malformations, and irregularities of the vessels, it must not, however, be supposed that these anomalies are to be regarded as distinct and independent deviations from the natural process of development. On the contrary, as the position of the heart varies with the different stages in the evolution of the organ and its connexion with the primary vessels, arrest or perversion of the process of growth may occur coincidentally and produce similar irregularities in all these particulars. Such is not, however, by any means always the case; in many

instances the heart being misplaced, the organ itself imperfectly formed, or the vessels arising from it irregular, without any deviation from the natural evolution of the other
parts. It is thus convenient to treat of these several defects
separately.

In applying any system of classification to individual
cases, difficulties will, however, often occur. For to
decide the position which any given case of anomaly
should occupy, it is necessary to ascertain what has been
the primary defect. This is often very difficult when the
specimen can itself be examined, and becomes almost impossible in published cases, which are often imperfectly and
sometimes incorrectly described. In the following pages I
cannot venture to hope that these difficulties have always
been satisfactorily solved. I have indeed been often compelled to refer to the same case as illustrating different
forms of defect, or to speak of anomalies which may only
have been secondary or necessary results of other changes
which could not be ascertained. This objection applies,
however, not only to the arrangement here adopted, but
would equally occur in any other which has been proposed.
I have also, in some cases, intentionally deviated from the
method of arrangement laid down. Thus, under the heads
of atresia of the pulmonic orifice and of constriction at the
point of union of the sinus with the infundibular portion of
the right ventricle, I have treated of all the forms of
anomaly, without reference to the period of foetal life at
which, as indicated by the state of the inter-ventricular
septum, the deviation occurs. To have classed these cases
under different heads might have been more scientifically

accurate, but would certainly have rendered the subject very complicated and much less easily intelligible. I have also, when treating of the defects of the pulmonary valves which prevent the closure of the foramen ovale, alluded to the cases in which the valves are found diseased without any other imperfection in the conformation of the heart. The latter cases should more properly have been placed under the head of congenital diseases of the valves, but I have preferred the arrangement adopted from the close analogy which exists between them and the other forms of anomaly. I have only to add, that it has not been thought necessary generally to quote more recently published cases, or to add to the number of those which had previously been subjected to statistical analysis. To have done so would have occupied considerable additional space, and have entailed much labour without any proportionate advantage. Where, however, recently reported cases have either modified views previously expressed, confirmed inferences which were before doubtful, or thrown light upon forms of anomaly imperfectly understood, I have not failed to refer to them or quote them more in detail.

TABLE OF CONTENTS.

TABLE OF CASES.

Fig. 2.

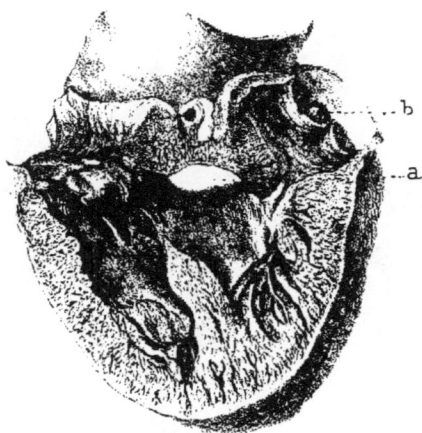

PLATE I.—Seat and Form of Apertures in the Septum of the Ventricles.

Fig. 1. Drawing of the left ventricle of the heart in the case exhibited at the Pathological Society by Dr. Quain; removed from a youth 18 years of age; referred to at p. 31.

Fig. 2. Drawing of portions of the heart in the case of Dr. Oldham—described at pp. 32 and 56.

    a. Exhibiting the aperture in the septum of the ventricles as viewed from the right ventricle, and showing that the opening is into the *sinus* of the right ventricle.

    b. Displays the form of the valvular apparatus of the pulmonary artery. The curtains are only two in number, but the larger valve shows the remains of division, in the form of a frenum or band extending from the edge of the valve to the side of the vessel.

The preparation of this heart is numbered B 11 in the Museum of the Hospital for Diseases of the Chest, Victoria Park. It was removed from a child 17 months old.

Fig. 1

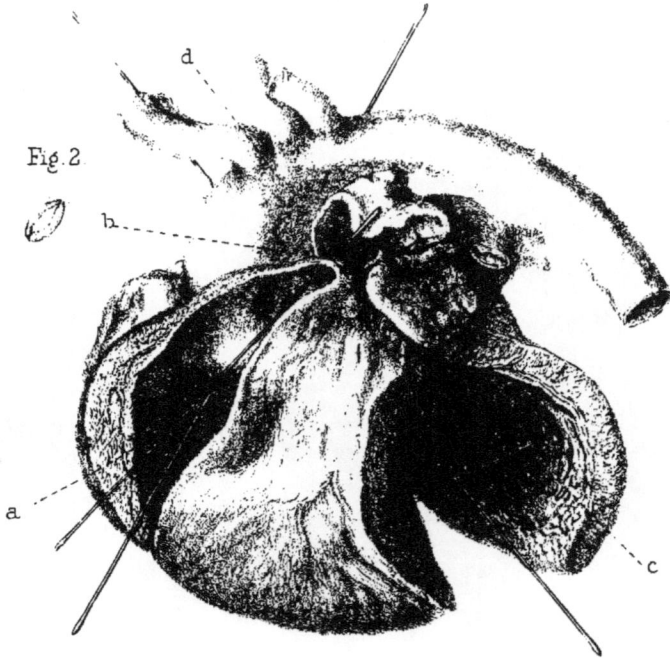

d

Fig. 2

b

a

c

Fig. 3

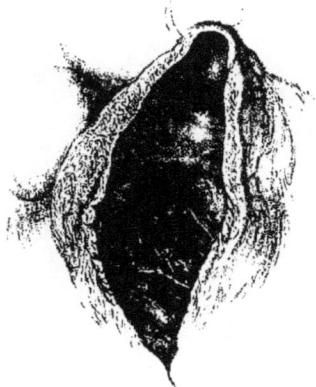

Fig. 4.

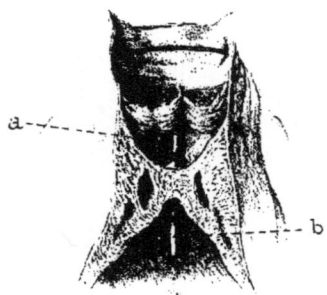

a

b

E. Burgess del et lith

W. West imp

PLATE II.—Constriction at the origin of the Pulmonary Artery, and defect in the Septum Ventriculorum, with the Aorta arising in part from the Right Ventricle.

Fig. 1. Drawing of the heart in Case I., described at p. 45.
The preparation is marked B 4 in the Museum of the Victoria Park Hospital. The child was 2 years and 5 months old.
   a. The right ventricle laid open.
   b. The contracted aperture of the pulmonary artery.
   c. The left ventricle.
   d. The ascending aorta.
The bristles passed into the aorta from behind, and visible above the upper edge of the vessel, are seen to enter both ventricles.

Fig. 2. Form of the valves of the pulmonary artery in this specimen.

Fig. 3. Front view of the conus arteriosus, or infundibular portion of the right ventricle in this case, to show the diminution in its size, and the mode in which it terminates in the protrusion forward of the valve and the slit which forms the pulmonary orifice.

Fig. 4. Diagram to show the extreme constriction at the commencement of the conus arteriosus, the atrophy of that portion of the ventricle, and the form of the arterial opening in Case VIII., p. 84. The pulmonary valves are only two in number, and one of them displays appearances of imperfect division. The subject of the disease was a boy 7 years old.
   a. Rudimentary conus arteriosus.
   b. Piece of wood passed through opening in septum between the conus and the sinus of the right ventricle.

PLATE III.

Fig. 2.

Fig. 1.

Fig. 3.

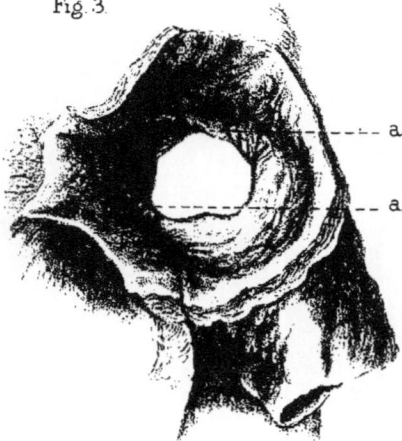

Fig. 4.

Fig 5.

T. West & E B del. E. Burgess lith   W West imp

PLATE III.—Constriction at the Orifice of the Pulmonary Artery from adhesion of the Valves, with the Foramen Ovale widely open.

Figs. 1 and 2. Drawings of portions of the heart in Case XIII., described at p. 112.

The preparation is numbered B 3 in the Museum of the Victoria Park Hospital. The septum of the ventricles is entire, yet the hypertrophy of the right ventricle is seen to be very great. The young man who was the subject of the disease died at the age of 20.

    a. The right ventricle laid open to show the great hypertrophy of its parietes.

    b. The pulmonic orifice.

Fig. 2. The orifice of the pulmonary artery as seen from above.

The union of the three valves into one, the frena or bands which mark the imperfect division of the segments, the thickening of the whole of the valves, and the form of the orifice, are well shown in this drawing. The aperture is seen also to have been permanently patent.

Fig. 3. The foramen ovale in the same case, showing that the process of closure has never been completed, the cornua of the valve, *a, a,* still remaining widely apart.

Figs. 4 and 5. Form of valvular apparatus in Case III., p. 51. From a boy who died at the age of 6½. The preparation is numbered B 6 in the Museum of the Victoria Park Hospital. In the second figure the valves are drawn as seen from above; the first is a diagram of their supposed appearance on division.

Fig. 1.

Fig 2.

Fig 3.

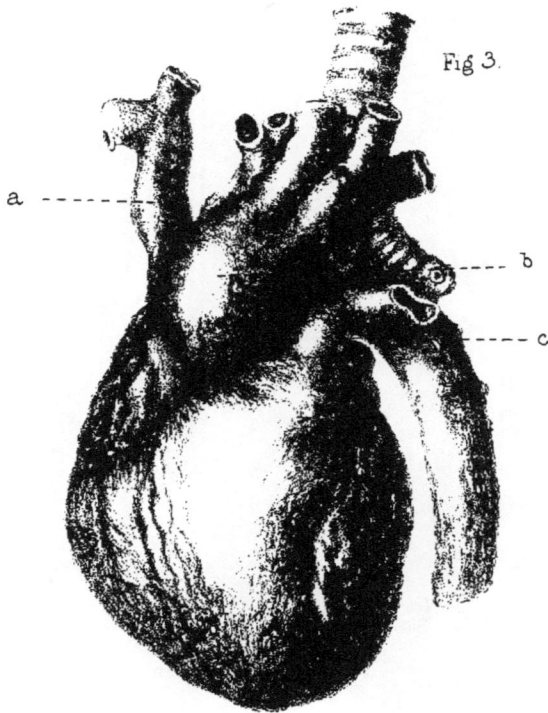

a -----

b -----

c -----

T. West del. E. Burgess lith.   W West imp

PLATE IV.—Forms of obstruction at the Orifice of the Pulmonary Artery, with irregular Course of the Aorta, etc.

Fig. 1. Form of the valvular apparatus in the specimen referred to at p. 44, existing in the Museum of St. Thomas's Hospital, and numbered LL 70. It has probably been removed from a young person 10 or 12 years of age.

Fig. 2. Drawing of obstruction at the outlet of the right ventricle, as existing in Case IV., described at p. 53.

The preparation is marked B 7 in the Victoria Park Museum. It was removed from a girl 19 years of age.

The valves are seen to be two in number, but one of them presents the appearances of former division. The orifice is of considerable size, but the obstruction at the exit from the ventricle is extreme, as shown in the drawing.

Fig. 3. Drawing of the heart in Case XII., described at p. 102, in which the pulmonary artery was of small size, probably from premature obliteration of the arterial duct. The aorta arose in part from the right ventricle, crossed over the right bronchus, and then behind the trachea, and gave off four vessels—the right and left carotid and the right and left subclavian arteries. The specimen was removed from a boy 11½ months old. The preparation is marked B 19 in the Victoria Park Hospital Museum.

    a. Descending cava.
    b. Aorta.
    c. Pulmonary artery.

PLATE V.

Fig 1.

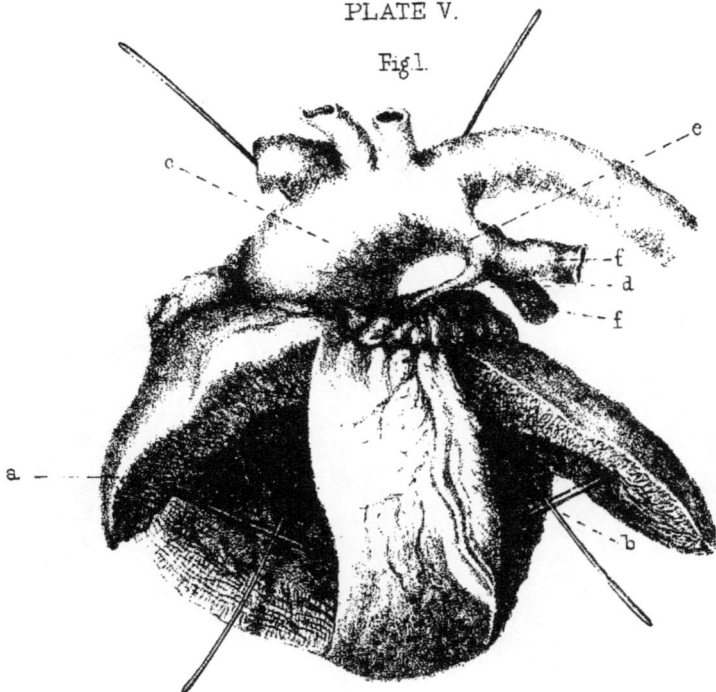

Fig 2.

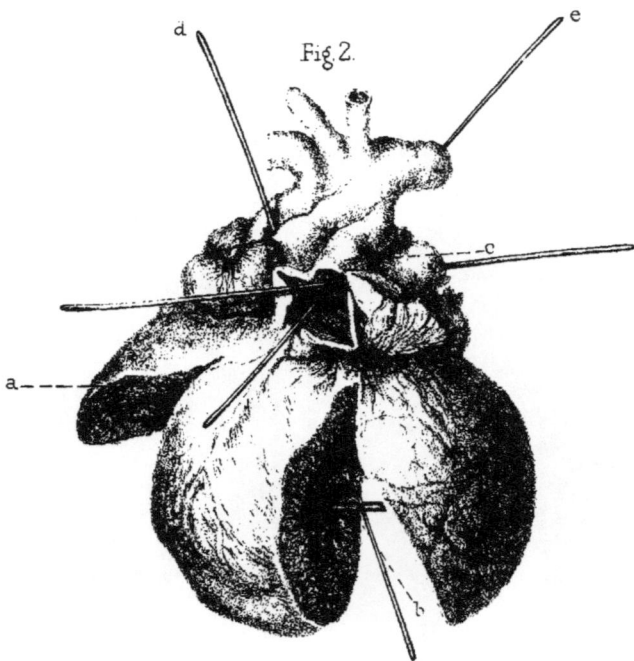

E Burgess lith ad nat   W West imp

PLATE V.—Obliteration of the Orifice and Trunk of the Pulmonary Artery; pulmonary branches supplied through the Ductus Arteriosus.

Fig. 1. Drawing of obliteration of orifice and trunk of pulmonary artery in Case V., described at p. 68.

The child which was the subject of the malformation died when nearly 1 year old, and the specimen is numbered B 8 in the Museum of the Victoria Park Hospital.

  a. Right ventricle.
  b. Left ventricle.
  c. Ascending aorta.
  d. The obliterated trunk of pulmonary artery.
  e. The open ductus arteriosus.
  f f. The pulmonary branches.

Fig. 2. Obliteration of orifice of pulmonary artery from adhesion of valves. Case VI., described at p. 72. The child only lived 9 days. The preparation is numbered B 25, Victoria Park Hospital Museum.

  a. Right ventricle.
  b. Left ventricle.
  c. Trunk of pulmonary artery.
  d. Probe passed into aorta protruding from left ventricle.
  e. Probe passed along pulmonary artery, showing the trunk of the vessel and ductus arteriosus to be largely open.

The septum of the ventricles is entire, and the small size of the right ventricle and the very large left ventricle are shown in the drawing.

Fig.1

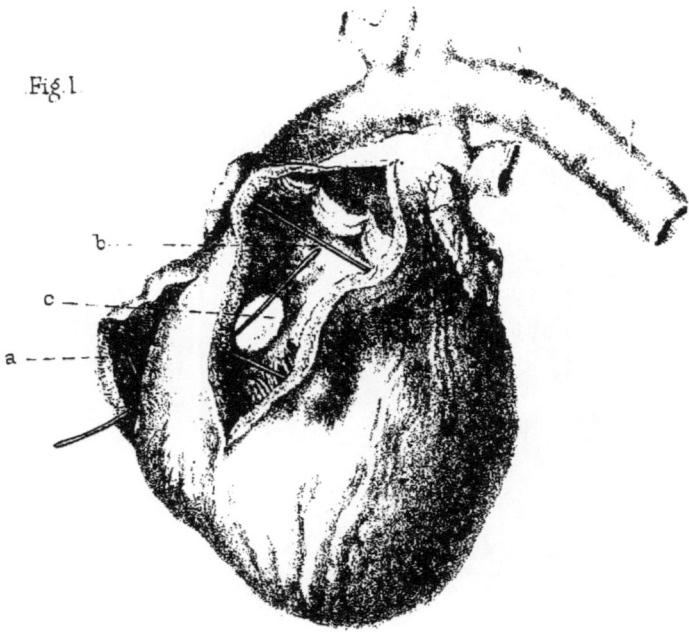

b .....

c .....

a .....

Fig.2

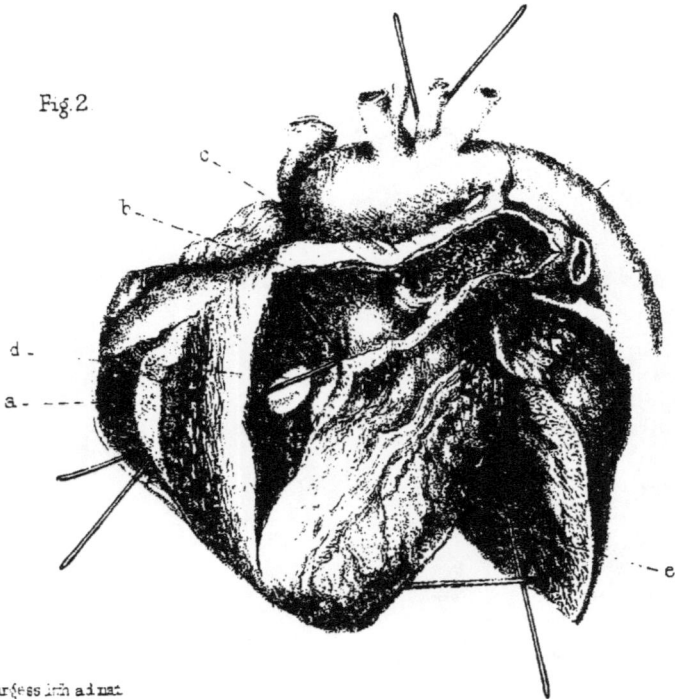

c .....

b .....

d .

a .....

e .....

Burgess lith ad nat.   W.West imp

PLATE VI.—Constriction between the sinus and infundibular portion of the Right Ventricle.

Fig. 1. Septum in the right ventricle, producing a marked separation between the sinus and infundibular portion. Heart otherwise well formed. Case IX., described at p. 86.

The child died of hæmorrhage during scarlatina, when 5 years of age.

The specimen is marked B 2 in the Victoria Park Hospital Museum.
   a. Sinus of right ventricle laid open.
   b. Infundibular portion of right ventricle.
   c. Opening in the septum between the two cavities.

Fig. 2. Partition between the two portions of the right ventricle with deficiency of the inter-ventricular septum and contraction of the pulmonic orifice. From a youth aged 15. Case VII., described at p. 79. The preparation is numbered B 5 in the Museum of the Victoria Park Hospital.
   a. Sinus of the ventricle with walls greatly hypertrophied.
   b. Infundibular portion with walls relatively thin.
   c. Valvular apparatus, and trunk of pulmonary artery obstructed by fibrinous coagula.
   d. Opening in the supernumerary septum.
   e. Left ventricle.

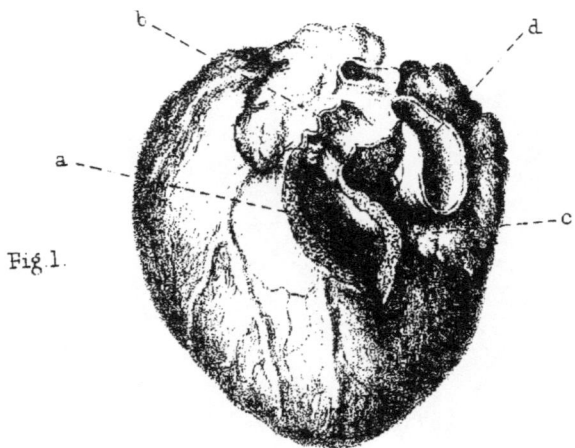

b　　　　　　d

a

c

Fig.1.

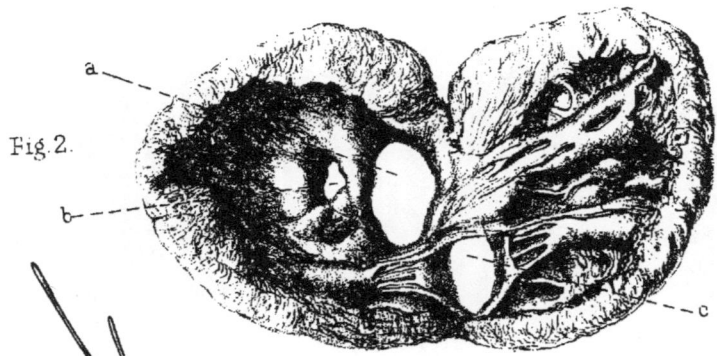

a

Fig.2.

b

c

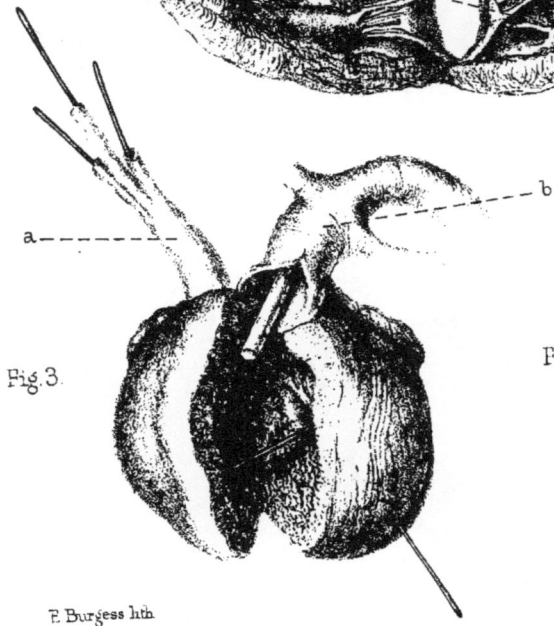

b

a

Fig.3.

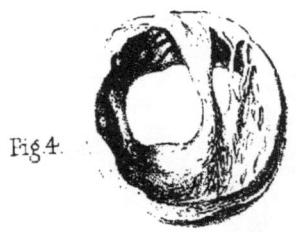

Fig.4.

P. Burgess lith.

W West imp.

PLATE VII.—Transposition of Aorta and Pulmonary Artery.
Descending Aorta derived from Pulmonary Artery.
Open Foramen Ovale.

Figs. 1. and 2. Drawings of transposition of aorta and pulmonary artery.
Case XVII., p. 147. From a child which survived 8 months.

Fig. 1. Shows the right side of the heart entire.
a. Small rudimentary right ventricle.
b. Aorta arising from that cavity.
c. The aperture by which the right ventricle communicated
with the large left ventricle.
d. Pulmonary artery.

Fig. 2. Cavity of the left ventricle laid open so as to show—
a. The origin of the pulmonary artery from that cavity.
b. The aperture which led into the small rudimentary right
ventricle.
c. The auriculo-ventricular aperture.

Fig. 3. Drawing of heart in case of Dr. Wale Hicks. Referred to at
p. 154. From an infant which only lived 13 hours.

a. The ascending aorta arising from the left ventricle, and
giving off the branches to the head and upper extremities.
b. The pulmonary artery forming the descending aorta.

Fig. 4. Open foramen ovale in case referred to at p. 115. The preparation
was removed from a girl aged 8, and is numbered B 16 in the
Museum of the Victoria Park Hospital.

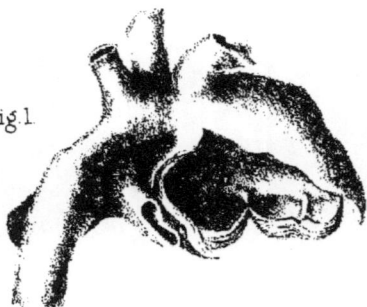

Fig. 1.

Fig. 2.

Fig. 3.

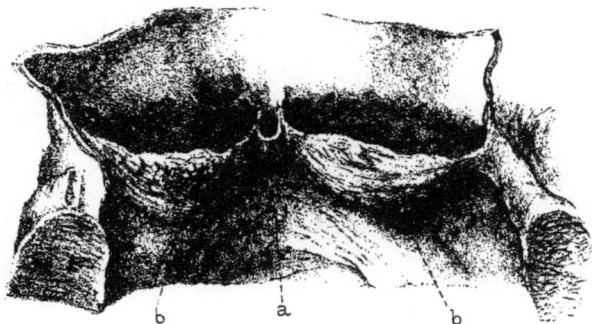

Fig. 4.

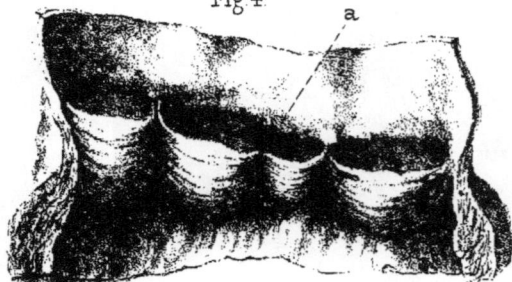

Fig. 5.

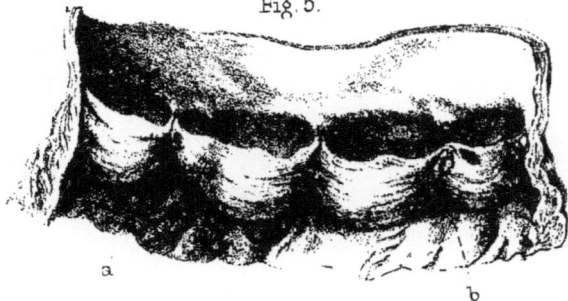

PLATE VIII.—Descending Aorta partly derived from the Pulmonary Artery. Defect and excess in the number of the Semilunar Valves.

Fig. 1. Fusion of two of the aortic valves, from a child 10 weeks old, whose case is described at p. 152, Case XVIII. There was some contraction of the aorta distal to the left subclavian artery and an open ductus arteriosus. The preparation is marked B 17 in the Museum, Victoria Park Hospital.

Fig. 2. Fusion of two of the aortic valves. Drawing of a specimen removed from a boy, aged 15, who was crushed to death.

It is numbered B 14 in the Museum of the Victoria Park Hospital.
In both these specimens the frenum or band, marking the former point of union or the imperfect division, is well seen in the larger valve.

Fig. 3. Drawing exhibiting the form of defect, in which one valve, *a*, has become atrophied from disease in fœtal or early life ; the larger valves, *b, b*, also exhibit the effects of subsequent disease so often seen in these cases. From a preparation in St. Thomas's Hospital Museum, removed from a man 60 years of age.

Fig. 4. Four valves at the orifice of the pulmonary artery, the excess being apparently produced by the division of one of the valves at *a*. The two segments so produced are imperfect and are partly blended together. From a female 75 years of age. The preparation is numbered B 13 in the Museum of the Victoria Park Hospital.

Fig. 5. Five valves at the orifice of the pulmonary artery, from a preparation marked B 12 in the Museum of the Victoria Park Hospital, removed from a child aged 4½ years. The excess is apparently due to the division of two curtains at *a* and *b*. The supernumerary segments and those adjacent to them are imperfect.

I.

MISPLACEMENTS OF THE HEART.

THE heart may be congenitally misplaced in various ways, occupying either an unusual position within the thorax, or being situated external to that cavity.

INTERNAL MISPLACEMENTS—ECTOCARDIA INTRA-THORACICA.

Transposition—Dexiocardia.

The most frequent of the internal misplacements is that in which the heart is placed in a position on the right side of the chest corresponding to that which it should occupy on the left. When this occurs, the viscera of the body generally are also most usually transposed; but such is not always the case, the heart being occasionally situated on the right side, while the other viscera retain their natural positions. The occurrence of misplacement of the heart to the right side in conjunction with transposition of the viscera generally, has been long known occasionally to occur, cases of the kind having been met with in Rome in 1643,[1] in Paris in 1650,[2] and in London in 1674;[3] and numerous ex-

---

[1] & [2] Thomas Bartholinus, Hist. Anat. Cent. ii. Hist. 29. Amstelodami, 1654, f. 199. In the man who was executed for the murder of the Duke of Beaufort, as related by Guy Patin. The case seen at Rome by Servius, and one by Schenkius, are also referred to.

[3] Phil. Trans. 1674, vol. x. No. 107, p. 146.—See also Riolanus, Op. Anat. 1649,—Sandifort, Obs. Anat. Path., lib. i. cap. ii. p. 39.—De Blegny, Zodiacus Medico-Gallicus. Genevæ, 1680, p. 129.

amples of the same condition have since been placed on record.   The occurrence of transposition of the heart while the other organs retain their natural positions, is of much more rare occurrence; but M. Breschet in his memoir[1] states that he has seen four instances of this in newly-born children; and Otto also refers to similar cases.[2]

When the heart is transposed, it may be well formed, as in the instances related by Dr. Sampson,[3] M. Méry and M. Morand,[4] Dr. Baillie,[5] M. J. F. Meckel,[6] M. Dubled,[7] M. Bosc,[8] and Dr. Allen Thompson.[9] In cases of this kind the vessels may retain their natural relations, or the arteries may be misplaced relatively to the ventricles, as in a case reported by Mr. Gamage,[10] in which the aorta arose from the right ventricle, and the pulmonary arteries from the left.   In other cases the heart may be very imperfectly developed, and the vessels arising from its cavities or entering into them, may be irregular, as in the instances related by Breschet,[11] Valleix,[12] Martin,[13] and Boyer.[14]

In cases of transposition, the aorta generally follows an irregular course, crossing over the right bronchus, and pass-

---

[1] Sur l'Ectopie de l'appareil de la Circulation et particulièrement sur celle du Cœur; Rép. Gén. d'Anat. et de Phys. Pathol., t. ii. 1826, p. 1.

[2] Selt. Brob. pt. 1, p. 95, and pt. ii. p. 47.

[3] Phil. Trans. for the year 1674, vol. ix. No. 107, p. 146.

[4] Hist. de l'Acad. Royale des Sc., t. ii. 1686 to 1699; Paris, 1733, p. 44; observation contributed in 1688.

[5] Phil. Trans., 1788, pt. 1, p. 350; Works by Wardrop, vol. i. p. 148.

[6] De Cordis conditionibus abnormibus, Dissertatio inauguralis, Halæ, 1802, Tab. 1 and 4.

[7] Arch. Gén. de Méd. 2mo année, 1824, p. 573.

[8] Bullet. de la Soc. Anat. de Paris, an. 4, 1829, p. 42.

[9] Glasgow Medical Journal, vol. i. 1854, p. 216. In a man forty-eight years of age, who died of broncho-pneumonia. The viscera generally were displaced.

[10] New England Journal of Medicine and Surgery, vol. iv. 1815, p. 244. See also a paper on transposition of the heart and viscera, by Valleix. Bullet. de la Soc. Anat. de Paris, 9me année, 1823, p. 253.

[11] Op. Cit., 1re obs. p. 7.

[12] Bullet. de la Soc. Anat., 9me 10me année, 1834–35, p. 253.

[13] Breschet, Op. Cit., 2me obs. p. 9; and Bullet. de la Soc. Anat. 1826, p. 39.

[14] Arch. Gén. de Méd., 4me série, t. 23, 1850, p. 90; and Gaz. Méd. de Paris, 20me an. t. 5, 1850, p. 292.

ing to the right side of the bodies of the vertebræ, and the right carotid and subclavian arteries are given off as separate trunks, while the brachio-cephalic trunk is placed on the left side. This occurred in a case described by Mr. Abernethy, in an infant of ten months, which has been frequently referred to as having presented an unusual arrangement of the vessels of the liver—the portal veins having terminated in the vena cava, and the liver being supplied with blood from an unusually large hepatic artery.[1] Such an arrangement does not, however, always obtain. In some instances, as in the case related by Mr. Douglas Fox,[2] the aorta, after crossing the right bronchus, passes behind the lower end of the trachea over the bodies of the vertebræ, and pursues its usual course to the left of the spine. In some instances, also, though the aorta may occupy the right side of the spine, the vessels at the arch are not transposed, the brachio-cephalic trunk being situated on the right side, and the left subclavian and carotid arteries arising separately.

Though a considerable number of instances of transposition of the heart have been observed, I have not myself had the opportunity of examining any case of the kind after death. In 1849, however, a boy presented himself at the Victoria Park Hospital for Diseases of the Chest, in whom the heart was placed on the right side, its apex beating an inch and a half below the right nipple, while the liver was situated on the left side. He had been an invalid since he was three years old, and, although then eighteen years of age, he looked like a much younger person, and had a peculiar sickly, unhealthy appearance; but there were no signs of any defect in the conformation of the heart.

More recently a patient was in St. Thomas's Hospital, under the care of Dr. Bristowe, in whom the heart was obviously situated on the right side, and the liver on the

---

[1] Phil. Trans. 1793, p. 59.
[2] London Medical and Physical Journal, vol. li. Jan. to June, 1824, p. 474. In a fœtus of the fifth month.

left.   The patient, a female, twenty-two years of age, was
labouring under phthisis.[1]

Mesocardia.

When the heart is seriously malformed, it is frequently
also found to retain the situation in the median line of
the chest, which it occupies at the earlier periods of fœtal
life.   Of this form of misplacement M. Breschet gives an
instance, which will again be referred to as an example of
defective development of the organ.   It is also mentioned
as existing in a case recently related by M. Kussmaul; and
I have myself seen the same anomaly under similar circumstances.   M. Breschet also admits the existence of dis-
placements in the *transverse* and *antero-posterior directions*,
but gives no examples of either of these forms.

EXTERNAL MISPLACEMENTS—ECTOCARDIA EXTRA-
THORACICA.

The most remarkable deviations from the natural position
of the heart are, however, those in which the organ is situated
wholly or in part external to the thoracic cavity.   Of this
form three varieties have been described by Breschet.

Ectocardia Pectoralis.

(*Ectopia Pectoralis Cordis, Weese and Breschet.*)—This
form of misplacement may occur without any defect or fissure of the thoracic parietes or with such imperfections.
As examples, Breschet refers to cases by Shulz and Vau-
bonnais, in which the misplacement existed in fœtuses; to
others, by Buttner and Weese, in a child which lived thirty-
six to forty hours; by Martinez, in which life was prolonged
twelve hours; and by Sandifort, in which the child survived

---

[1] Similar cases are quoted in the Bull. de l'Acad. de Méd., t. xxvi. 1860–61,
p. 1174, in a female of twenty-one, still alive; and in the Lancet for 1863, by
Dr. Maclean, vol. ii. p. 159, in a soldier of twenty-five years of age.

for a day. Alone, this form of malformation is rare; but when the viscera of the abdomen also are protruded, it is not of uncommon occurrence. Cases have more recently been related by Dr. O'Bryan,[1] MM. Cruveilhier and Monod,[2] Mr. Mitchell,[3] Mr. Sydney Jones,[4] and Mr. Daniel.[5]

Ectocardia Abdominalis.

(*Ectopia Cordis Ventralis, Weese and Breschet.*)—Of this form of misplacement there are also two varieties. In one, the heart is protruded through the diaphragm without forming a tumour externally; in the other, there is an external tumour. As examples of the first kind, M. Breschet refers to the case of Ramel, in which a female, ten years of age, had the heart situated immediately below the diaphragm. A much more remarkable instance is that related by M. Deschamps of Laval, of a man who had served in the army, and retired in consequence of suffering from severe pains in the loins. He, however, married, and had three children; but the lumbar pains continued, he became emaciated, and died exhausted by continued suffering. On examination, the right kidney was found large, hard, and in a state of suppuration; and a solid mass containing a cavity filled with sanies was situated in the pelvis. The heart occupied the place of the right kidney, and the vessels arising from it passed through an opening in the diaphragm into the thorax.

Of the second variety, the case of Mr. Wilson to be hereafter mentioned, and others by Prochaska, Klein, Sandifort, Chaussier,[6] &c., afford examples; and cases have recently been published by M. Follin[7] and Mr. Barrett.[8]

---

[1] Prov. Med. and Surg. Trans., vol. vi. 1837, p. 374.
[2] Gaz. Méd. de Paris, 2me série, 9mo an. 1841, p. 497.
[3] Dublin Journal, vol. xxvi. 1844, p. 262.
[4] Path. Trans., vol. vi. 1854-5, p. 98.
[5] Brit. Med. Journ. 1860, p. 776.
[6] Bull. de la Soc. de Méd., t. iv. 1814, p. 93.
[7] Arch. Gén. de Méd., 4me série, t. xxiv. 1850, p. 101; and Gaz. Méd. de Paris, 3me série, t. v. 1850, p. 629.
[8] Lancet, 1834 and 1835, vol. i. p. 349.

Ectocardia Cervicalis.

(*Ectopia Cordis Cephalica, Breschet.*)—In this form of
displacement, the heart lies in the front of the neck, in
connexion with the ramus of the jaw. In the only cases on
record, the malformation existed in fœtuses or in infants
which scarcely survived birth, and which presented other
serious defects. It is therefore much less important than
the other forms.

The most interesting cases of this description of misplace-
ment which have been published since the appearance of
M. Breschet's memoir are those of M. Cruveilhier and M.
Monod, Dr. O'Bryan and Mr. Sydney Jones, and Mr. Daniel.

In the case of M. Cruveilhier and M. Monod, the child
lived only a few hours. The heart lay externally, having
escaped through a round opening in the upper part of the
sternum, and was exposed as if after the pericardium had
been incised.

Mr. Sydney Jones's case occurred in the practice of Dr.
Bain of Poplar, and the specimen was exhibited at the
Pathological Society in 1855. The child was believed to
have been born at the eighth month, and it survived thirteen hours. The heart, devoid of pericardium, was situated
wholly external to the thoracic cavity. A fissure existed in
the sternum of an oval shape, its longer diameter being
nearly vertical and three-quarters of an inch in length, and
its shorter diameter transverse and half an inch long.
Through this opening the vessels passed from the heart into
the thorax. The margin of the fissure was obscured anteriorly by a prolongation of skin on the great vessels, and
from thence on the external surface of the heart. The
cuticle could be traced as far as the base of the viscus and
slightly over the auricles; but beyond this point there was
no epithelial covering,—the muscular substance being only
protected by a structure shown on microscopical examination to be a white fibrous, mixed with yellow, elastic tissue. The closing up of the foramen was completed poste-

riorly by the reflexion of the pleura from the great vessels on to the parietes of the thorax. The heart appears to have been well formed, and the child presented no other defect.[1]

In the case of Mr. Daniel of Newport-Pagnell, the child did not breathe after birth, but the heart continued to beat for four hours. It was situated in the front of the chest, having protruded through an aperture formed by deficiency of the middle and lower part of the sternum. The septum of the ventricle was defective. The other organs of the body occupied their natural positions.

In these cases the heart lay in front of the chest, and they therefore constituted examples of the Ectopia Pectoralis of Breschet; but in the cases of Dr. O'Bryan and Mr. Barrett the heart was situated partly in the abdomen, having passed through an aperture in the diaphragm. They were therefore examples of ectopia ventralis.

In Dr. O'Bryan's case, the xyphoid cartilage was deficient, together with the fibres of the diaphragm inserted into that body; a triangular aperture was thus left, through which a portion of the left ventricle, covered with pericardium, was protruded, so as to form a soft, oval, unequal and semi-transparent tumour at the anterior and upper part of the abdomen. The part of the left ventricle protruded was one inch and three-quarters in length, and a further portion was prevented escaping by the apex of the right ventricle. The lower part of the tumour was occupied by a portion of colon. The child survived three months, and died of bronchitis brought on by exposure.

In Mr. Barrett's case, the organ was situated partly in the abdomen, the anterior portion of the diaphragm being deficient. The arrangement of the primary vessels was irregular, the aorta being derived chiefly from the pulmonary artery. There was also an umbilical hernia, and the child was still-born.

---

[1] The preparation is contained in the Pathological Museum at St. Thomas's Hospital, and is marked LL 56.

In the Museum of St. Thomas's Hospital there is a specimen which affords an example of partial ectocardia abdominalis. There is an aperture in the diaphragm three inches in diameter, at the edges of which the pericardium and peritoneum are continuous, and through this opening the apex and lower part of the heart protrudes, so that the organ is situated partly in the pericardial and partly in the peritoneal cavity. The heart rests upon the left lobe of the liver, which to some extent closes the defect in the diaphragm. A large portion of the great omentum is situated within the pericardial sac, and is attached by adhesions to the lower half of the anterior surface of the heart, and to the corresponding part of the reflected pericardium. Considerable portions of the heart and of the orifice are free from attachments, so that the finger might be readily passed from the pericardium through the opening to the upper surface of the left lobe of the liver. The preparation was removed from the body of a man, a labourer, forty-seven years of age, who died of gangrene and inflammation of the lung, with chronic ulceration of the fauces.[1]

---

[1] LL 80.

For reference to various cases of misplacement of the heart, see Otto's Pathological Anatomy (South's translation), pp. 274-6 ; and Förster's Missbildungen des Menschen. Jena, 1861, taf. xviii. figs. 5 to 7.

II.

DEFICIENCY, ETC. OF THE PERICARDIUM.

INSTANCES of the asserted absence of the pericardium have not unfrequently been recorded; yet it is probable that in some of these cases the membrane has not been really wanting but only universally adherent, and that true congenital absence of the pericardium is of rare occurrence. When, however, the heart is much misplaced, it is also very generally deprived of its pericardial covering, as in some of the instances before referred to ;[1] and occasionally the sac is wanting, when the organ occupies its natural position and is otherwise well formed.

One of the earliest authentic examples of this defect is that which occurred to Dr. Baillie,[2] in 1778 ; though that writer refers to cases related by Columbus, Bartholinus, and Littre, and to one mentioned in the Philosophical Transactions for 1740.[3] Dr. Baillie states that, on opening the cavity of the chest of a man 40 years of age, who had not presented any symptoms referable to the heart, in order to explain to his class the situation of the thoracic viscera, he was exceedingly surprised to find the heart lying naked in the left side. It was bare and distinct, and lay loose in the left cavity of the chest, unconnected in any way, except by its vessels. It was somewhat large, and was also placed

---

[1] This is shown in the specimen of biloculate heart, described by Mr. Wilson, to be hereafter referred to, of which the preparation is contained in the Museum of the Royal College of Physicians, and a specimen in St. Thomas's Museum, LL 80.

[2] Transactions of a Society for the Improvement of Medical and Surgical Knowledge, vol. i. 1791, p. 91 ; and Works by Wardrop, vol. i. p. 44.

[3] See also references in Otto's Pathological Anatomy (South's translation), p. 254, sects. 3 and 4.

at a lower level than usual.   The mediastinum consisted of
two layers of pleura united by cellular tissue.   The right
phrenic nerve ran between the laminæ of the mediastinum,
near the right side of the heart, and the left phrenic nerve
was situated between the same layers, almost immediately
under the sternum.

In 1826, M. Breschet[1] described a similar defect, which
was found in a man 28 years of age, who died of acute
dysentery, under the care of M. Pettit.   He had previously
enjoyed good health.   The heart was found lying loose in
the left pleural sac, except that there existed adhesions
between the base of the left ventricle and the left lung, and
the apex of the heart and the diaphragm.   The specimen
was exhibited at the Académie de Médecine, and was examined by Cruveilhier, Laennec, Blainville, &c.

In 1839, a similar malformation fell under the notice of
Mr. T. B. Curling,[2] in the body of a man aged 46, who died
of paraplegia, at the London Hospital.   He was not known
to have manifested any peculiarity in the circulation.   The
heart occupied its natural position.   There were some
white spots on its surface, and the corresponding portions of
the pleura covering the lungs were opaque and thickened,
and the lower lobe of the left lung was adherent to the heart.

In 1851, Dr. Baly[3] exhibited at the Pathological Society,
a specimen removed from a man 32 years of age, who died
of phthisis in the Millbank Penitentiary.   He had never
exhibited any symptoms of "obstructed or disordered circulation."   The heart had no separate sac, and was in contact
with the left lung, but not attached to the diaphragm.   The
left side of the heart and the left lung displayed some

---

[1] Sur un Vice de Conformation Congéniale des Envelopes du Cœur.   Rép.
d'Anat. et de Phys. Path., t. i. p. 67.   A good drawing is given of this specimen,
in which the crescentic fold constituting the rudimentary pericardium is well
shown, and the left phrenic nerve is seen to pass to the right of the heart beneath
the serous covering on the anterior surface of the fold.

[2] Med.-Chir. Trans., vol. xxii. N. S. vol. iv. p. 22.   Mr. Curling also refers to
a case related in Rust's Magazine, not here quoted, vol. xxxiii. p. 333.

[3] Path. Trans., vol. iii. 1850-51, 51-52, p. 60.

recent false membrane, and there was a slight adhesion between the apex of the heart and the lung. The left pleura was reflected over the heart and vessels, and then passed forwards to the sternum, and was there separated from the corresponding portion of the right pleura only by a thick layer of fibrous and cellular tissue. There was a rudimentary pericardium, in the form of a crescentic fold of the serous membrane, reflected beneath and behind the heart. This " fold, thickened by fibrous tissue between its layers, arose on the right side of the ascending aorta, passed downwards to the' right of the right auricle, and in front of the inferior vena cava, and crossing behind the left auricle, terminated in the left pulmonary veins." Since the exhibition of the last specimen, a second has been shown at the Pathological Society by Dr. Bristowe. It was removed from the body of a man 28 years of age, who died of disease of the mitral valve, with pulmonary congestion and jaundice. The heart was considerably enlarged, and both auricles and the right ventricle were greatly dilated. The heart and left lung were both contained in the left pleural sac, and the lower part of the upper lobe of the lung was firmly attached to the anterior surface and left side of the heart, and to the left auricle. A fold of membrane, or as it is termed by Dr. Bristowe, " a diverticulum, or pocket," exists at the upper part and right side of the heart, which is evidently the rudiment of the pericardium. It commences on the pulmonary artery, passes over the aorta and vena cava descendens to the diaphragm and ascending cava, and thence it is continued to the left till it is lost in the bronchi and vessels at the root of the lung and in the left auricle. It consists of fibrous tissue, covered on each side by pleura, and where widest, at the side of the right auricle, is about an inch and a half in depth. The fold is adherent to the heart in several places. The right phrenic nerve took its usual course, but the left nerve passed vertically downwards beneath the serous membrane[1] of the crescentic fold, about

---

[1] Path. Trans. vol. vi. 1854–55, p. 109. The specimen is contained in St. Thomas's Museum, and is marked LL. 78.

half an inch from its edge. The membrane representing
the rudimentary pericardium, mentioned as existing in these
cases, was also noticed in those of M. Breschet and Mr.
Curling. In all the cases, the left phrenic nerve passed to
the right of the heart, either beneath the outer serous covering of the rudimentary membrane, or between the layers of
pleura forming the mediastinum.

In Dr. Baly's and Dr. Bristowe's cases, I had the opportunity of examining the specimens, and, some time previously, an instance of partial deficiency of the pericardium
fell under my own notice. The heart was separated from
the right pleural sac by a layer of serous membrane covering
a fibrous septum, while on the left side it lay loose in the
cavity of the pleura, in connexion with the lung, with which
it had contracted adhesions. From the circumstances under
which the post-mortem examination was performed, I was
not able to examine the parts so carefully as would have
been desirable. The subject of the case was a man 75
years of age, who died of aortic valvular disease. In several
of the other cases mentioned the heart was healthy.

A specimen existing in the Museum at St. Thomas's
Hospital, which exhibits a partial defect in the pericardium
with displacement of the heart, has been before mentioned.
In this instance, the base of the pericardium with the
corresponding portion of the diaphragm is deficient, so that
the cavities of the pericardium and peritoneum freely
communicate.[1]

I do not know that the mode in which the pericardium
is developed has been made the subject of investigation, but
from examining the adult heart it would appear to be a
continuation of the fibrous sheath of the vessels to the
diaphragm and over the heart. When the membrane is
fully developed, and the layers passing in front and behind
the heart come in contact on the left side and become adherent, the sacs of the pleura and pericardium will be distinct; but if the growth be arrested, so that the two layers

---

[1] LL 80.

do not become united, the heart will lie in the pleural cavity, and the pericardium will only be represented by the crescentic fold, consisting of fibrous tissue covered on each side by pleura, which has been noticed as existing at the right side of the heart in nearly all the recorded cases. That this is the correct explanation of this form of defect, is confirmed by the fact that when the heart is seriously misplaced, being situated externally at front of the chest, the organ is very generally deprived of pericardium. In M. Breschet's case, it is stated that the rudimentary membrane was not attached to the diaphragm, the defect being apparently at the base of the pericardium. This will probably be the case generally, when, from defect in the diaphragm, the heart is situated in the abdomen, and under these circumstances also the heart is usually devoid of pericardium.

A preparation is contained in the Museum of St. Thomas's, which exhibits a diverticulum of the pericardium, but it is doubtful whether this is to be regarded as an instance of redundant development, or as the result of disease in after-life. The diverticulum communicates with the common cavity by a small aperture and a tubular passage about an inch in length. The cyst itself is about an inch in diameter, and contained limpid fluid.[1] Its coats are thin and transparent, and it seems probable that it may have originated in distension of a part of the pericardium by fluid. A specimen which may probably have been very similar to this, was exhibited by Mr. Hird at the Westminster Medical Society in 1848.[2]

---

[1] LL 79.

[2] Lancet, vol. ii. p. 64, from a man sixty-five years of age, in whom there was valvular disease and lymph on the pericardium.

III.

MALFORMATIONS OF THE HEART.

I. MALFORMATIONS CONSISTING IN ARREST OF DEVELOPMENT OCCURRING AT AN EARLY PERIOD OF FŒTAL LIFE.

HEART CONSISTING OF TWO CAVITIES.

HEARTS consisting of only two cavities, an auricle and a ventricle, with a single vessel supplying both the systemic and pulmonic circulations, have, though rarely, been found in infants which have survived for a short period after birth.[1] If we except the cases of Pozzi[2] and Lanzoni,[3] which are too imperfectly described to be relied upon, the first instance of this description of malformation was related to the Royal Society by Mr. Wilson in 1798.[4] Since that time an example of biloculate heart occurred to Dr. Farre in 1814,[5] and a specimen was presented to the Pathological Society by Mr. Foster in 1846.[6] A heart exhibiting a somewhat more advanced degree of development was described by Mr. Standert,[7] in the Philosophical Transactions for 1805, and a similar case was met with at Philadelphia, by Mr. Mauran, in 1827,[8] and others have since been described

---

[1] Otto, in his Pathological Anatomy (South's translation, pp. 269, 270), refers to cases in which, in imperfectly formed fœtuses, the heart consisted only of an expanded vascular trunk or a single valveless cavity.

[2] Miscellanea Curiosa, Med. Phys., sive Eph. Med. Phys. Germ., annus quartus et quintus anni 1673–74, Francofurti et Lipsiæ, 1676, obs. 40, p. 37.

[3] Ibid. Norimbergæ, 1691, obs. 44, p. 79.

[4] Phil. Trans., vol. lxxxviii. 1798, p. 346.

[5] On Malformations of the Human Heart, 1814, p. 2.

[6] Path. Trans., vol. i. 1846–47, 1847–8, p. 48.

[7] P. 228. See also Ramsbotham, in Path. Trans., vol. i. p. 48.

[8] Arch. Gén. de Méd., t. xix. p. 257, quoted from Philadelphia Journal of Med. and Phys. Sc., vol. xiv. 1827 ; N. S. vol. v. p. 253.

by M. Thore,[1] Dr. Crisp,[2] Professor Owen and Mr. Clark,[3] and Dr. Vernon.[4]

The subject of the case recorded by Mr. Wilson was a child born at the full period, and which lived seven days. Though occasionally livid it was generally pale, and died of sloughing of the parietes of a sac which contained the heart, and which, owing to a considerable defect in the tendinous portion of the diaphragm and to the absence of the lower part of the pericardium, formed a tumour projecting from below the sternum to the middle of the abdomen. Within this sac the heart rested upon the convex surface of the liver. The organ consisted of a single auricle and ventricle, and gave off a vessel which divided into two branches; one of these furnished the pulmonary arteries, while the other proceeded upwards behind the thymus gland, and then gave off the usual aortic vessels. The thymus was unusually large, the pulmonary artery was much smaller than the aorta, and there do not appear to have been any traces of the ductus arteriosus. There were only two pulmonary veins, and these entered the descending vena cava. There were neither bronchial arteries nor veins.[5]

In Dr. Farre's case, the child at birth was of full size, and though it breathed with some difficulty, and was slightly livid, it subsequently, for forty-eight hours, seemed to enjoy perfect health. " His countenance was lively and ruddy, his skin warm, and he took the breast eagerly." His breathing then became difficult and remarkably quick, the heart beat strongly, and his cries expressed distress. The skin became pallid and cold, the pulse at the wrist could not be felt, and he died seventy-nine hours after birth. On examination, the heart consisted of only one auricle, ventricle, and artery. The venæ cavæ opened into the auricle,

---

[1] Arch. Gén. de Méd., 3me et Nouvelle Série, t. xv. 1842, p. 316.
[2] Path. Trans., vol. i. 1846-7, 1847–48, p. 49.
[3] Lancet, 1848, vol. ii. p. 664.
[4] Med.-Chir. Trans., vol. xxxix. 1856, p. 300.
[5] This specimen is retained in Dr. Baillie's Museum, now in the possession of the Royal College of Physicians. It is numbered 4 A, 17.

and the pulmonary veins into the appendix, which was more distinctly separated from the sinus than usual. There was only one ostium ventriculi, and the ventricle gave origin to a single vessel, which furnished, first the two pulmonary branches, and then the usual systemic arteries, together with a vessel which passed down to the heart, and formed the coronary arteries.

The case of Mr. Forster was very similar to the last. There was only one auricle, which received the two cavæ and two pulmonary veins, and opened into a ventricle from which a single vessel originated. This artery gave off two pulmonary branches, and the coronary arteries were derived from a trunk which apparently arose from the concavity of the aortic arch. The subject of the case was a male infant to all appearance well developed; it however refused the breast, and had several attacks of dyspnœa, in one of which it died seventy-eight hours after birth.[1]

In each of these cases the condition of the heart was very rudimentary, and the peculiar origin of the coronary arteries would appear to indicate that the division of the primary vessel had not taken place. In the following cases the heart had attained a higher stage of development.

In the case which occurred in the practice of Mr. C. Clark and is described by Professor Owen, the heart consisted of three cavities, two auricles with distinct auricular appendages, and one ventricle; but the left auriculo-ventricular aperture was obliterated, so that the left auricle communicated with the common ventricle only through the medium of the foramen ovale and the right auricle. The ventricle gave origin to a single artery, which gave off the pulmonary arteries separately, and the coronary arteries arose by a common trunk from the innominata. The child in which this malformation was found was puny and livid when born, but acquired a more natural colour afterwards. It again became livid, and died convulsed on the third day.

---

[1] A heart exhibited at the Société Anatomique, by M. Gibert, appears to have been of this description, but it is very imperfectly described. 2me série, t. v. 1860, p. 55.

The subject of Dr. Vernon's case was a robust male infant, which appeared healthy at the time of birth, but on the second day its breathing became difficult and the surface discoloured, and it died in convulsions on the third day. On examination, the right auricle was found of large size, and was separated from a rudimentary left auricle only by a fleshy column and the fold of the valve. The left auricle received two pulmonary veins, but had no direct communication with the ventricle. The right auricle opened into a ventricle which was without any division, and gave origin to a single vessel. This vessel first gave off two pulmonary arteries, and then divided into the arteria innominata and the left carotid and subclavian arteries. The coronary arteries were derived from a common trunk which proceeded from the innominate artery.

The case related by Mr. Standert, in the Philosophical Transactions, occurred in the practice of Dr. Combe, and the preparation, which is contained in the collection of Dr. Ramsbotham, was exhibited at the Pathological Society in 1846, and is re-described in the first volume of the Transactions.[1] Through the kindness of Dr. Ramsbotham, I have had the opportunity of examining this specimen. There are two distinct auricular appendages, and the division of the cavity is indicated by a muscular band in the usual situation of the septum. The ventricle is large, and of a somewhat quadrangular form, and gives origin to the aorta, from which the coronary arteries arise as usual. There is another cavity forming a cul-de-sac in front of the aortic orifice, which is partially separated from the larger ventricle, and is evidently the analogue of the right ventricle, though it does not present any rudiment of a pulmonary artery. From the published account it appears that the pulmonary circulation was supplied from the aorta through the ductus arteriosus ; but that vessel is not retained in the preparation. We are informed that the child from which the heart was removed lived ten days, and was very livid throughout its life, though

---

[1] Page 48.

C

the functions of respiration and nutrition appeared to be otherwise naturally performed.

The cases of Mr. Mauran, M. Thore, and Dr. Crisp afford examples of still more advanced development. In the first of these the auricle was provided with two distinct auricular appendages. The cavæ entered on the right side of the common cavity, and the pulmonary veins, which were only two in number, on the left. The auricle opened into an undivided ventricle, by an aperture guarded by a valve having the tricuspid form. The ventricle gave origin to the aorta, and also to a rudimentary pulmonary artery which was obliterated at its commencement. The trunk of the latter vessel, however, was pervious, and the pulmonary branches had received a supply of blood from the aorta, through a largely-open ductus arteriosus. In this case the child, which was a female, at the time of birth was small, but appeared healthy. When moved, it had attacks of difficulty of breathing in which it became livid, and uttered cries of distress, and it died in one of these when ten and a half months old.

In 1842, M. Thore described the case of a female infant which he saw at the Hospice des Enfans Trouvés, which presented general cyanosis, and had attacks of dyspnœa, with rapid breathing, a dry cough, and slight convulsions. It died when somewhat more than four months old. The heart consisted of only two cavities—an auricle and a ventricle; the former was nearly spherical, and appears to have had two imperfectly developed appendages; it received the systemic and pulmonic veins, and opened into the ventricle by a single aperture. The ventricle gave origin both to the aorta and pulmonary artery; but the latter vessel was much the smaller of the two. The ductus arteriosus did not exist.

The case of Dr. Crisp was that of a child which had the usual symptoms of blue disease, and which died convulsed when ten weeks old. The preparation is retained by Dr. Crisp, and I have been favoured by that gentleman with the opportunity of examining it. The heart affords a good example of the transition from a single to a double set of cavities. There are two auricles, of which the sinus and appendix of the right are of large size, while the left auricle is rudimentary, and the inter-auricular septum is very imperfect. The auricles receive the usual veins, and open by a common aperture into a large ventricular cavity—the right ventricle—which is in connexion with a rudimentary, but impervious, pulmonary artery. From the upper and right side of this cavity there is a communication with a second smaller sac—the representative of the left ventricle, from which the aorta takes its origin. The aorta doubtless furnished the supply of blood to the pulmonary branches and lungs through the ductus arteriosus; but that vessel has been cut away in making the preparation.[1]

From the description which has been given of these cases, it will be evident that they afford examples of hearts in very different stages of development. The cases of Mr. Wilson, Dr. Farre, and Mr. Forster, presented the simple biloculate condition of the organ; while those related by Mr. Clark and Professor Owen, Dr. Vernon, Mr. Standert and Dr. Combe, Mr. Mauran, M. Thore and Dr. Crisp, illustrate the gradual advancement from the simple form of the heart to that in which it consists of four cavities. The cases of Mr. Clark and Professor Owen and Dr. Vernon are more closely allied to the former class of malformations; the others more nearly approximate to that which is next to be described.

More recently a case has been described by M. Claude Bernard which affords an example of imperfect development of the heart closely allied to those which have been just described. The child lived about three weeks. The auricle had two imperfectly formed appendages, but there was not even the rudiment of a septum, and the single auriculo-ventricular aperture opened into a ventricle, evidently the left, from which the aorta arose. At the anterior part of this

---

[1] A case very similar to some of those which have been quoted, is referred to by Dr. Chevers, as seen in an infant which lived nine days, by M. F. Tiedemann.

cavity there was a small cavity the analogue of the right ventricle, which had, however, no connexion with the auricle,
was not larger than a nut, and did not give origin to
any vessel. The source of the pulmonary supply and the
relations of the systemic and pulmonary veins with the
common auricle are not described.[1] This case bears a close
analogy to that described by Dr. Crisp, except that while in
M. Bernard's case the fully developed ventricle was the left,
in that of Dr. Crisp it was the right. The defects in these
cases might probably originate in obliteration, at early
periods of fœtal life, of one or other of the auriculo-ventricular apertures.

In the volume of the Pathological Transactions for 1864-
65, a specimen is described which was exhibited by Mr. C.
Heath for Mr. Power, and which I have since had the
opportunity of examining.[2] It was removed from the body of
a child which lived only twenty-four hours, and was also the
subject of encephalocele. The heart gave origin to only one
vessel, which arose from the right ventricle and gave off first
the pulmonary arteries, and then, while making its turn, the
usual branches at the arch. The two auricles were distinct,
but the foramen ovale was not entirely closed. The left auricle
received as usual the pulmonary veins and opened into a
small cavity—the analogue of the left ventricle, and this
cavity had no artery arising from it, but opened by a small
deficiency in the septum into the right ventricle. At
the base of the heart there was a small vessel which had
been cut across and was apparently the coronary artery.
This must have derived its origin, either from the common
trunk at a much higher point than natural, or from one of
its branches. In this case, therefore, though the heart
itself had undergone more complete development than in
any of the cases last named, the separation of the primitive
vessel into the aorta and pulmonary artery had apparently
not taken place.

---

[1] L'Union Médicale, Nouvelle Série, t. v. 1860, p. 612.
[2] Vol. xvi. p. 62.

HEART CONSISTING OF THREE CAVITIES.

In this description of malformation, the process of development is not arrested till a later period of fœtal life than in the class of cases last named, and the organ is found to have attained a more advanced condition. Thus, the auricular sinuses are separated by a more or less complete septum, and there are generally two auriculo-ventricular apertures ; while the ventricle is either wholly undivided or presents only a very rudimentary septum. The arteries which are given off are usually two in number,—an aorta and pulmonary artery. This kind of defect, though very rare, is of more frequent occurrence than the biloculate heart. A case, which appears to have been of this description, was described by Chemineau, in 1699 ;[1] one was related by Tiedemann, in 1808-10 :[2] a specimen exhibiting a similar state was shown to Dr. Farre by Mr. Lawrence, in 1814,[3] and cases have been described by Fleischmann in 1815,[4] by Hein in 1816,[5] by Kreyzig[6] and Wolf in 1817, and more recently by Breschet,[7] Thore,[8] and Hale.[9]

In the case of Chemineau, the child had apparently only just breathed, and the heart is stated to have had three ventricles, one of which received the vena cava, another the pulmonary veins, and both opened into the third, from which the aorta and pulmonary artery arose. The pulmonary artery was small, but gave off the branches to the lungs. There was no communication between the pulmonary artery and descending aorta.

Tiedemann's patient, who had suffered from the usual

---

[1] Hist. de l'Acad. Royale des. Sc., année 1699, p. 37.

[2] Zoologie, t. i. p. 177.   [3] Malformations, p. 30.

[4] Meckel, Arch. f. d. Phys., 1815, p. 284.

[5] De istis cordis deformationibus quæ sanguinem venosam cum arterioso misceri permittunt. —Gœttingæ, 1816, p. 37.

[6] Krankheiten des Herzens, vol. iii. p. 200.

[7] Sur l'Ectopie, etc., obs. 1.

[8] Arch. Gén. de Méd., t. i. 4me série, 1843, p. 199.

[9] Path. Trans., vol. iv. 1852-53, p. 87.

symptoms of morbus cæruleus, lived to the age of eleven, and the heart was found to have two auricles and one ventricle, and from the latter cavity the aorta and pulmonary artery arose.

In the specimen which was shown to Dr. Farre by Mr. Lawrence, the venæ cavæ and pulmonary veins opened as usual into their respective auricles ; but the inter-auricular septum was very imperfect, consisting only of a small muscular band, which left a large foramen ovale without any valve; the septum ventriculorum was altogether wanting, and the ventricles communicated with the auricles by a common aperture. The aorta and pulmonary artery arose from the left side of the ventricle, and the orifice of the latter vessel was somewhat contracted. The history of the child was not known ; but, from the size of the heart, it was inferred to have lived some months.

The case of Fleischmann differed in some degree from these, as, though the heart consisted of three cavities, the ventricle only gave origin to one vessel, the orifice of the pulmonary artery being impervious. The child had lived twenty-one weeks. Hein's patient was a young man, who had been livid and had suffered from difficulty of breathing and other symptoms of malformation from early life, and who died of an abscess in the lung when sixteen years of age. There were two auricles, but the fold of the foramen ovale was perforated in three places. The auricles led by distinct apertures into a ventricle which had only the vestige of a septum, in the form of a membranous fold extending from the apex on the posterior side. The aorta and pulmonary artery arose from the common ventricle, and the latter vessel was of small size and its valves converted into a ring. The ductus arteriosus was closed.

The case described by Kreyzig, is one which occurred to Mr. Wolf, and the heart is preserved in the Berlin Museum. It consists of two well-formed auricles with distinct auriculoventricular apertures. There is an opening in the valve of the foramen ovale, and only one ventricle from which two arteries arise. The valves of the pulmonary artery are imperfect. The subject of the malformation was a young man twenty-three years of age.

In the Archives Générales de Médecine for 1828,[1] a case is quoted from Hufeland's Journal, which occurred to M. Wittcke, in a man twenty-four years of age, who had been subject to violent palpitation from infancy. He was attacked with peripneumony, followed by aggravated difficulty of breathing, amounting ultimately to orthopnœa; dropsical symptoms set in, and he died exhausted. The pericardium was adherent to the heart everywhere except at the apex, where it was wanting. The heart was very greatly enlarged, and the walls of the ventricle were fully three times their usual thickness : there was not a trace of the interventricular septum, but the positions of the vessels of the heart were natural and the orifices were somewhat dilated. A volume of the same journal for the previous year[2] contains a case quoted from a German publication, which may have been similar to this, but it is very imperfectly reported. The patient had always been delicate and livid, had suffered from spitting of blood, palpitation, and dyspnœa, and died of marasmus when thirty-five years old. It is stated that there was no inter-ventricular septum; but in this there seems some mistake, as the cavity of the right ventricle is reported to have been almost obliterated by the thickening of its parietes.

M. Breschet, in 1826, described the heart of a child which he had examined with M. J. F. Meckel of Halle, and which presented this form of malformation. The child was a male, and was born at the full period, did not present any peculiarity of colour, and lived for one month. In addition to other imperfections, a spina bifida, &c., the heart was situated on the right side of the chest. The right auricle was very large, and had two openings by which it communicated with the left auricle. There were two descending venæ cavæ, and the pulmonary veins opened into the right auricle. The ventricle was undivided, and gave origin to a large aorta

---

[1] T. xviii. 6me année, 1828, p. 83.   [2] T. xv. 5me année, 1827, p. 110.

which passed to the right side of the spine. The pulmonary artery was imperforate at its orifice, and its branches received their blood through the ductus arteriosus.

In 1853, Dr. Hale exhibited to the Pathological Society the heart of a male infant, which when born was healthy-looking and well formed, but had afterwards occasional attacks of vomiting with difficulty of breathing coming on in paroxysms. While quiet the surface was warm, but during the attacks the extremities became cold. When seen by Dr. Hale, ten weeks after its birth, the surface of the body and the lips and hands were pale, but had the natural tint. The pulsations of the heart were tumultuous and strong, and the pulse full and rapid. There was a superficial whizzing sound, heard at the sternum near the third cartilage. The child survived nine weeks longer, and was found dead in bed. The heart was much enlarged. The auricles were dilated, the appendices much increased in size, and the lining membrane thickened. The foramen ovale was wide and open, and the valve-like fold thicker than usual. The tricuspid orifice and valves were natural, but the mitral orifice was large and patulous. The ventricle was without the slightest rudiment of a septum. The aorta and pulmonary artery arose in their ordinary positions : the former was about double its natural size, the latter being much smaller than usual. The semilunar valves of the pulmonary artery were large, thickened, and irregular, but apparently competent : the aortic valves were natural. The ductus arteriosus probably existed, though it did not occupy its usual situation.

The cases which have just been quoted corresponded so far as that the heart consisted of two auricles and only one ventricle ; but in other respects they differed in some degree. Thus, in the cases of Fleischmann and M. Breschet, the common ventricle gave origin to only one vessel—the aorta ; while the pulmonary artery, though it existed, was impervious at its orifice, and its branches received the blood which they transmitted to the lungs, from the aorta through the ductus arteriosus. In all the other cases, on the contrary, there were two distinct vessels given off from the common ventricle. In the case of Mr. Lawrence, also, there was only one auriculo-ventricular aperture; whereas, in the other cases, two distinct openings existed.

In some cases which are on record, the heart has been found to present this form of malformation, combined however with irregularity in the origin of the large vessels. M. Breschet[1] has described the case of a male infant, whose heart was exhibited at the Société Anatomique,[2] by M. Martin. The child was born at the full period, and lived six weeks, suffering during that time from dyspnœa, vomiting, and convulsions. It was cold but not apparently cyanosed. The heart was situated in the median line, with its apex slightly inclined to the left side; it was of natural size, and had only one auricle, but with distinct auricular appendages; and there were two ascending and two descending cavæ. The aorta and pulmonary artery arose from the common ventricle, but their points of origin were transposed. There were two communicating arteries in the place of the ductus arteriosus;—one of which united the brachio-cephalic trunk and the right branch of the pulmonary artery; the other passed from the left pulmonary branch to the aorta.

A heart, displaying defects in some respects similar to those in the last case, has been described by M. Thore.[3] There were two distinct auricular appendages, but the cavities of the auricles freely communicated. The ventricle also presented the rudiments of a septum, but there was only one auriculo-ventricular aperture, and the points of origin of the aorta and pulmonary artery were transposed. The infant which was the subject of the malformation lived eleven days. It had a dry, hard cough, and dyspnœa increased on drinking liquids; but it was not cyanotic.

In the Museum of St. Thomas's Hospital there is a preparation which affords another example of this condition.

[1] Sur l'Ectopie, obs. 2me, p. 9.   [2] Bulletin, 1re année, 1826, p. 39.
[3] Arch. Gén. de Méd., 4me série, t. i. 1843, p. 199.

The specimen has no history attached to it, having been obtained in the dissecting-room; but, from the size of the heart, it was probably removed from a child eight or ten years of age. There are two auricles; but the right auriculo-ventricular aperture is obliterated, so that the blood from the right cavity must have flowed through a largely open foramen ovale into the left auricle, and thence into the ventricle. The auriculo-ventricular valve rather resembles the mitral than the tricuspid valve. The ventricle is a single cavity, but presents a rudimentary septum, in the form of a thick fleshy column extending down its posterior wall, and the two arteries to which it gives origin are transposed;—the aorta arising in front, in the usual situation of the pulmonary artery, while the latter vessel proceeds from the posterior part of the ventricle.[1] Cases of this description are closely allied to those hereafter to be described, in which, with transposition of the vessels, the septum of the ventricles is defective, and they only differ from them in the extent of the septal imperfection.

II. MALFORMATIONS CONSISTING IN ARREST OF DEVELOPMENT OCCURRING AT A MORE ADVANCED PERIOD OF FŒTAL LIFE.

HEART CONSISTING OF FOUR CAVITIES;—ONE OR BOTH OF THE SEPTA IMPERFECT—PULMONARY ARTERY AND AORTA MORE OR LESS COMPLETELY DEVELOPED.

When the partition of the ventricle is incomplete the arrest of development may be almost entire, so that, as in some of the cases last mentioned, there may be merely a muscular band projecting into the cavity; or the deficiency may be very slight, one or more small apertures only existing at the upper part of the septum.

If the inter-ventricular septum be only partially defective, the imperfection most generally occurs at the base,

---

[1] LL 64.

where, during fœtal life, the division of the cavities is last effected. In this situation there naturally exists in the fully developed organ, a triangular space, in which the ventricles are only separated by the endocardium and fibrous tissue on the left side, and by the lining membrane and a thin layer of muscular substance on the right. This space indicates the point at which, in the turtle, there is a permanent communication between the two aortic ventricles; and it is interposed in man between the base of the left and the sinus of the right ventricle. Laterally it is bounded by the attachments of the right and posterior aortic valves, and its base is formed by the muscular substance of the septum. The dimensions of the space vary with the size of the heart; but ordinarily in the adult, the sides may be estimated at about seven lines, and the base is somewhat wider. When the lower part of the space is perforated, the left ventricle and origin of the aorta communicate with the sinus of the right ventricle; but if the defect be situated high up, towards the angle of attachment of the valves, the communication may be between the left ventricle and the right auricle.

Generally when an aperture occurs in this situation, or, as it has been termed, in the *undefended space*, it has a triangular form; but, in some cases, it is oval or rounded; and in others, there are two or three apertures. When there is no source of obstruction at the right side, so that the right and left ventricles retain their just proportions, the flow of blood is from the latter cavity to the former, and the openings from the left ventricle are usually larger than those into the right ventricle. Not unfrequently, also, under such circumstances, as the apertures enter the right ventricle immediately below the ring of the auriculo-ventricular opening, the folds of the tricuspid valve become expanded by the column of blood flowing into the right ventricle, so as to form one or more small sacs. This has been pointed out by Dr. Thurnam[1] as shown in a specimen in the Museum of

---

[1] On Aneurisms of the Heart.—Med.-Chir. Trans., vol. xxi. 1833.

the Royal College of Surgeons, and probably the sac was thus formed in a case described by Dr. Pereira,[1] as one of partial aneurism of the heart.

While, however, deficiencies in the septum cordis are most commonly situated at the base, they are not confined to that situation. Occasionally, though, so far as my observation serves me, very rarely the division between the left ventricle and the infundibular portion of the right is perforated, so as to form a communication between the left ventricle and the origin of the pulmonary artery.[2] In some cases, also, perforations occur nearer the apex ; and in yet others, several openings have been found in different parts of the septum.

The defects in the inter-ventricular septum which have been enumerated, may co-exist with imperfect separation of the auricles, but such is not always the case ; and, on the other hand, the auricular septum may be imperfect or the foramen ovale unclosed, while the partition of the ventricles is entire.   When also the auricular cavities are imperfectly divided, the septum may be scarcely developed, and the valve entirely absent ; or the septum and valve may both exist, but only in a rudimentary form ; or the septum of the auricles may be fully formed, but the valve may be very imperfect, so as to be incapable of covering the opening ; or lastly, the orifice may be entirely closed by the valve, and an aperture may exist in some other part of the auricular partition.

Occasionally the imperfection exists at the base of the heart where the septa of the auricles and ventricles should unite, and thus a peculiar form of malformation results in which the four cavities communicate.   A case of this kind has been described by M. Thibert,[3] which occurred in a man who lived to the age of twenty-four, and presented no signs of disease of the heart till six weeks before his death.   A

---

[1] Lond. Med. Gaz., 1845.
[2] Two specimens illustrating this condition are contained in the Museum at St. Thomas's Hospital, LL 65 and 66.
[3] Journal Gén. de Méd., 2me série, t. viii. 1819, p. 254.

similar preparation was some time ago sent to me by
Dr. Curtis, of Alton, which had been removed from the
body of a girl, about twelve years old, of whose previous
history no satisfactory account could be obtained. In this
case the free communication between the left ventricle and
the right ventricle and auricle, had been prevented by the
expansion of a portion of the curtain of the tricuspid valve
in the manner just described.[1]

It has been contended by some writers, and especially by
M. Bouillaud, that the apertures which are found in the
septum ventriculorum are not always congenital; and there
can be no doubt that perforations of the septum do occasionally, though I believe rarely, take place as the result of
disease in after-life. I have myself met with two cases in
which disease existed at the undefended space at the base of
the left ventricle,[2] which would have led to perforation, had
life been prolonged; and a specimen of a similar description
was exhibited at one of the earlier meetings of the Pathological Society, by the late Mr. Avery. A case is also related
by M. Corvisart,[3] in which the writer was doubtful whether
the perforation was not due to disease; and Laennec has referred to a case in which he supposed an opening which
existed was the result of ulceration. I cannot, however,
agree with M. Bouillaud in regarding all the instances which
he has quoted, as affording examples of communications so
produced; and it is probable that some other cases, reported
by more recent writers as the results of disease, have been
congenital defects. Generally there can be but little difficulty in deciding as to the congenital or accidental origin
of the apertures in any given instance; for, in a large pro-

---

[1] Path. Trans., vol. i. 1846–47, 1847–48, p. 61.
[2] Path. Trans., vol. ii. 1848–49, 1849–50, p. 49. See also notices of perforations supposed to depend on disease, by Dr. Bennett, in vol. i. p. 59; and by
Dr. Wilks, in vol. vi. p. 103. In the Archiv. Gén. de Méd., 5me série, t. vi.
1855, p. 101, a case by M. Buhl is quoted from Henle's Zeitschrift, in which
an aperture was found in the septum of the ventricles in a female nineteen years
of age, which was ascribed to ulceration.
[3] Mal. du Cœur, 3me éd. 1818, p. 287.

portion of cases of defective development, the existence of other irregularities will afford unmistakable evidence of their nature; and, in others, the triangular or rounded form of the openings and their smooth and polished edges will be equally conclusive. It must also be borne in mind, that when the edges of the openings are rough and irregular from fibrinous deposits, affording evidences of endocarditis, this does not alone show that the apertures are the result of disease; for such changes constantly accompany defects unquestionably of congenital origin.

In 1865, Dr. Hare exhibited at the Pathological Society a very curious specimen, in which the base of the inter-ventricular septum at the undefended space was expanded so as to form a small sac projecting into the right ventricle;[1] and when in Vienna, in 1864, several similar specimens were shown me by Rokitansky. In cases of this kind, the membrane may readily give way in the process of expansion, and so an aperture of communication between the two sides of the heart be produced, which would very closely simulate a congenital defect.

Deficiencies in the partitions of the auricles and ven-tricles usually coexist with other important deviations from the natural form and development of the heart, and espe-cially with some source of obstruction at the pulmonic or other orifice. But even large apertures are occasionally found in one or other of the septa, in persons who have died of affections unconnected with the heart, who had pre-sented during life no signs of obstruction or disorder of the circulatory system, and in whom the organ is otherwise well formed and free from disease.

In the Museum of St. Thomas's Hospital there are spe-cimens which illustrate several of these forms of defective development. In LL 68 there is a small aperture in the undefended space of the left ventricle, by which that cavity communicates with the sinus of the right ventricle. This

[1] Pathological Transactions, vol. xvi. p. 81. The specimen was removed from the body of a man twenty-four years of age, who died of diphtheria.

specimen was removed from a man, twenty-three years of age, who died of phthisis, and does not appear to have presented any signs of cardiac lesion. LL 62 is a preparation of the heart of a child, six years of age, who presented no signs of disease of the heart till she had scarlet fever, ten weeks before death. When admitted into the hospital, two weeks after the commencement of her illness, she had palpitation, and a systolic bruit was heard over the whole chest, but loudest in the left axilla. She died of bronchitis and œdema of the lungs, and there were no indications of cyanosis. An opening, nearly circular in form and six lines in diameter, exists at the upper part of the septum atriorum above the foramen ovale, and the latter passage is closed except that a small valvular communication still exists. LL 69 exhibits a deficiency in the septum cordis of a young person. The aperture is of considerable size, and the aorta, which is unusually large, arises above it, so as to communicate with both ventricles. The fold of the foramen ovale is also unattached over a large space. The mouth of the pulmonary artery is not contracted.

LL 66 and 67 afford examples of the rare form of defect in which communications exist between the left ventricle and the infundibular portion of the right ventricle; but, as in these specimens there are other remarkable anomalies, I shall allude to them more fully hereafter.

A specimen in which a communication existed between the left ventricle and right auricle was exhibited by Mr. Daldy to the Hunterian Medical Society, in 1853, which I have been favoured by Mr. Hilton with the opportunity of examining.

In the first plate,[1] I have given a representation of deficiency in the inter-ventricular septum, from a specimen exhibited at the Pathological Society, by Dr. Quain, in 1856. The preparation was taken from a youth, eighteen years of age, who died of phthisis, at the Brompton Hospital, under the care of Dr. Cursham. He had been cyanotic since he

---

[1] Plate 1, fig. 1.

was two years of age.   The " aperture was sufficiently large
to permit a florin to pass through," and was remarkable for
occupying the whole of the undefended space, and for its
regular triangular form.   The other drawing,[1] in the same
plate, represents the usual situation of the opening into the
sinus of the right ventricle, and is taken from a specimen
exhibited by myself at the Society in 1849, and of which the
preparation is retained in the Museum of the Hospital for
Diseases of the Chest, Victoria Park.   It was removed from
a patient, whose case will be hereafter detailed, who had
presented the usual aspect of morbus cæruleus, and died
when seventeen months old.

DEFECT IN THE INTER-VENTRICULAR SEPTUM ; CONSTRICTION
OR OBLITERATION OF ORIFICES; MISPLACEMENT OF THE
PRIMARY VESSELS.

When the division of the ventricles is imperfect, there is
often some deviation of the septum, so that the origins
of the aorta and pulmonary artery are misplaced ; and such
deviation generally coexists with some source of obstruction
to the flow of blood from the right ventricle by the pulmonary artery ; more rarely with constriction of one of the
auriculo-ventricular orifices, or of the aortic aperture.

In one of the most interesting forms of anomaly the
deviation of the septum is to the left, so that the right
ventricle is of large size, and the aorta arises wholly or in
part from that cavity ; and this condition is most generally
associated with obstruction to the passage of the blood
from the right ventricle.   The source of obstruction in
such cases may be situated either—1st, at the outlet of the
pulmonary orifice, or at the free edge of the valves ;
2ndly, at the line of attachment of the valves to the fibrous
zone, or the termination of the infundibular portion of the
ventricle ; 3rdly, in the course of the pulmonary artery ;
or 4thly, at the commencement of the infundibular
portion of the right ventricle, or the point of union of the
infundibular portion and sinus.

---

[1] Plate 1, fig. 2.   The preparation is marked B 11 in the Museum.

Of these several forms of defect I shall proceed to speak, leaving for the present the consideration of those cases in which similar effects upon the development of the heart result from congenital smallness of the pulmonary artery.

,1. *Constriction or Stenosis*[1] *of the orifice of the pulmonary artery ; aorta arising in part from the right ventricle.*

The first reported case of this kind of malformation, is one which fell under the notice of Sandifort,[2] and was published in 1777. The subject of the case was a boy, aged twelve and a half years, who had experienced great difficulty of breathing and palpitation of the heart, and had been unusually livid since he was a year old. The pulmonary orifice, owing to the adhesion of its valves, was so contracted as only to admit the passage of a small probe. The septum of the ventricles was imperfect, so that the aorta arose in part from the right side, and the foramen ovale was also open. In 1783 Dr. Hunter reported a case which he had occasionally seen for several years,[3] and which proved fatal in 1761. The boy was always dark-coloured, and had presented the usual symptoms of malformation of the heart since shortly after birth, and was remarkably thin. He was liable to paroxysms of difficulty of breathing, but could arrest them by lying down on the carpet when they were coming on. There is some uncertainty as to the age to which this patient survived : in the paper he is stated to have been thirteen years old, but, in the description of the plate, he is said to have been only eleven. On examination, the orifice of the pulmonary artery was found very greatly contracted, and the septum of the ventricles was imperfect, as in the case of Sandifort ; but the foramen ovale, though not expressly mentioned, may be inferred to have been closed. In 1785 a third case of the kind was published by

---

[1] στενὸς, narrow.
[2] Obs. Anat. Pathol. Lugduni Batavorum, cap. i. 1777, fig. 1. Sandifort subsequently met with a similar case in dissecting the body of a fœtus.
[3] Med. Obs. and Enq., vol. vi. 1783, p. 299, case 2.

D

Dr. Pulteney,[1] which had occurred in 1781, and appears to
have been communicated to the College of Physicians
before the publication of Dr. Hunter's paper. The patient
was a boy who died of acute dysentery at the age of thirteen
years and nine months. The symptoms which he presented
were similar to those in the two former cases, and he was
especially remarkable for the darkness of his face and hands,
and his liability to faintness on exertion. The period at
which these symptoms first appeared is not mentioned. The
ring of the pulmonary artery was much smaller and firmer
than usual ; and the septum of the ventricles was defective,
so that the end of the finger could be passed from the aorta
into either ventricle. The foramen ovale was most probably
closed. In 1793, Mr. Abernethy[2] related an example of
this form of malformation, in which, however, the child died
when only two years old. The symptoms, as in the case
of Dr. Hunter, appeared shortly after birth. The pulmonary
artery was of small size ; the aorta arose from the right
ventricle ; the septum cordis was imperfect, and the foramen
ovale was largely open. In 1785, the history of a similar
case was communicated to Dr. Duncan[3] by Dr. Nevin of
Downpatrick. The child which was the subject of the
malformation presented no peculiarity till nearly two months
after birth ; it then had oppression of the chest, difficulty
of breathing, and lividity. The symptoms became more
marked when it was four months old, and it died at the age
of ten months. The aorta at its commencement was large,
and it was connected equally with both ventricles. The
pulmonary valves were adherent together and ossified at
their bases, and the artery was contracted. The foramen
ovale would admit a large probe, but the ductus arteriosus
was impervious. The case occurred at Glasgow, and the
specimen is said by Mr. Burns to be contained in Dr.

---

[1] Med. Trans. of College of Phys., vol. iii. 1785, p. 339.
[2] Med. and Surg. Rev., vol. i. 1795, p. 25 ; and Surg. Essays, 1793, vol. ii.
p. 157.
[3] Duncan's Medical Commentaries, Edinburgh, vol. xix. 1795, p. 325.

Jeffrey's Museum.[1] Since the publication of these cases many others have been described in this country, France, Germany, Italy, and the United States. Indeed, as remarked by Dr. Farre, of the various deviations from the natural conformation of the heart, defects of this kind are the most common.[2]

1. When, in cases of this description, the obstruction to the flow of blood from the right ventricle is due to disease of the pulmonic valves, the number of the segments is generally defective, and they are otherwise diseased. Thus, there may be only two segments; or the orifice may be imperfectly closed by a membrane stretched across and perforated in the centre; or the valvular apparatus may be still more imperfect, being only represented by a duplicature of the lining membrane, or by a band of muscular fibres surrounding the orifice.[3]

When only two valves exist, one of the curtains is usually larger than the other, and presents evidences that it originally consisted of two distinct segments, which have become united together; the former line of separation being indicated by a ridge or frenum, extending from the edge of the valve to the side of the artery. When the valvular apparatus has the form of a perforated membrane, or, as it has been termed, of a diaphragm, the three segments are united together, and there are three ridges or frena on the upper or arterial side, with corresponding sulci or furrows on the ventricular surface. Whatever be the form of the valvular apparatus, the curtains are generally protruded forwards in the course of the artery; so as to give the opening from the ventricle a funnel shape, and to form deep sinuses between the valves and the sides of the vessel. This protrusion is due to the constant pressure

---

[1] Burns on Diseases of the Heart, 1809, p. 13.

[2] A memoir on this form of malformation has recently been published by Dr. Mayer, of Zurich.—Virchow's Archiv, 12 Band, 1857, p. 497.

[3] Crampton : Trans. of College of Phys. of Dublin, N. S., vol. i. 1830, p. 34 ; and Cyclop. of Anat. and Phys., vol. ii. p. 634. Favell : Prov. Med. and Surg. Jour., vol. iii. 1842, p. 440.

exercised by the blood in the right ventricle on the obstructing membrane.

The orifice itself varies in shape according to the number of segments.   When there are two distinct valves, it is generally in the form of a slit, extending from side to side : when, on the contrary, there is only a perforated membrane, the opening is usually triangular or rounded.   In some cases, the bases as well as the free edges of the valves are contracted, and the orifice has a tubular or barrel shape.[1]

The valves are most commonly not only adherent, but thickened, irregular, and indurated ; and, if life be sufficiently prolonged, they not unfrequently contain larger or smaller masses of cretaceous deposit.   Generally, also, they become the seat of subsequent disease, and display recent deposits or vegetations, by which the contracted opening is often still further diminished in size.   In some cases, indeed, the obstruction is mainly due to warty growths from the valves.   Thus, in a case which occurred to Mr. Leadam, the orifice was so greatly contracted in this way as to be only large enough to admit a small probe.[2]   Usually, also, at least when all the segments are united together, the valves are incapable of entirely closing the aperture.

2. When the constriction is not caused by adhesion of the valves, but is situated at the outlet of the infundibular portion of the ventricle or conus arteriosus, it may depend on disease of the fibrous zone to which the valves are attached, or on hypertrophy of the adjacent muscular substance and thickening of the endocardium.[3]

3. Obstruction may also be due to disease of the pulmonary artery itself, that vessel being found contracted in its calibre and its coats thickened and indurated.

The size of the contracted orifice, whatever be the precise

---

[1] These different forms of valvular and other defects are figured in plate 1, fig. 2 ; plate 2, figs. 1, 2, 3, 4 ; plate 3, figs. 1, 2, 4, 5 ; plate 4, figs. 1, 2 ; plate 6, fig. 2 ; and plate 8, figs. 1, 2.

[2] Farre, p. 37.   [3] This condition is represented in plate 4, fig. 2.

seat of the constriction, varies greatly in different cases. It may admit the point of one of the fingers, or a lead pencil, or common quill; or it may only allow of the passage of a probe or crow-quill. M. Louis has recorded a case in which, in a man twenty-five years of age, it measured only two and a half lines in diameter (5·6 millimetres, ·133 English inches); and M. Bertin found the aperture of the same size in a female of fifty-seven. I have myself measured the capacity of the orifice in two males, fifteen and twenty years, and in a female nineteen years of age, and found the circumference thirteen (29·25 mm. 1·154 e. in.), twelve (27· mm. 1·065 e. in.), and eight (18· mm. ·71 e. in.) French lines respectively :—the average size in adults being about thirty-six French lines (81· mm. 3·19 e. in.). In four children, nine, seven, six and a half, and two and a half years of age, the circumference of the orifice was fifteen (33·75 mm. 1·332 e. in.), six and a half (14·62 mm.·577 e. in.), six (13·5 mm. ·532 e.in.), and five (11·25 mm. ·444 e. in.) French lines in circumference.

Whatever be the source of obstruction at the exit of the ventricle, the trunk of the pulmonary artery is most generally smaller than natural, and its coats thin and transparent, more resembling those of veins. Usually, however, the trunk of the vessel is much larger than would be expected from the small size of the aperture; and this, though the ductus arteriosus is obliterated, and there is no other means by which the blood could enter the artery. In some cases, the arterial coats are, as before said, thickened and indurated, the obstruction being situated in the vessel itself, or the artery may be involved in disease which may have been primarily seated in the valves.

In speaking of imperfection in the inter-ventricular septum, it has been mentioned that the defect may exist in very different degrees. The extent to which the septum may be misplaced may also vary greatly. In some cases only a small portion of the aortic orifice, as one-third or one-fourth of its circumference, is in connexion with the right ventricle. In others, the aorta is placed immediately

above the incomplete septum, so as to communicate equally
with both ventricles; and, in yet other instances, it arises
wholly from the right ventricle.   An example of the latter
condition was exhibited at the Pathological Society, in 1846,
by Mr. Ward and Dr. Parker.[1] The heart had been removed
from the body of a boy, who died of pneumonia, at the age
of thirteen, and had been cyanotic from birth.   The valves
of the pulmonary artery were united so as to form a
diaphragm, with an aperture in the centre which would only
admit a small quill.   The ascending aorta, as is always the
case when that vessel communicates freely with the right
ventricle, was of large size.   The left ventricle formed only
a small supplementary sac, opening from the right ventricle; and the latter cavity was unusually large, and its
walls thick.

In most cases of this description of malformation, the
foramen ovale does not become entirely closed; but such is
not always the case, the septum of the auricle being in some
instances completely imperforate; or the aperture closed by
the fold, though it does not become entirely adherent to
the isthmus.   In other cases, the foramen ovale may be
naturally occluded, but an aperture may exist in some
portion of the auricular partition.   The ductus arteriosus
occasionally remains pervious; and the tricuspid valves are
frequently diseased, the folds being adherent at their edges,
and their surfaces studded with fibrinous deposits or vegetations.

With an imperfectly divided ventricle, the heart may
possess its natural form externally, and the defect may only
be detected on laying open the cavity.   More generally,
however, the large size of the aorta and its situation too far
to the right, and the smallness of the pulmonary artery, are
at once apparent.   Usually, also, especially when the septal
defect is considerable, the organ is wider than natural,
resembling the quadrangular form of the heart in the turtle.
On closer examination, the infundibular portion of the right

---

[1] Path. Trans., vol. i. p. 51.

ventricle is found imperfectly developed, its cavity very small and the walls somewhat thick ; while the sinus is much enlarged, its parietes acquire a greatly increased width and are peculiarly tense and resisting, and the muscular columns become much hypertrophied. In a boy of ten, and a girl of nineteen years of years, the walls of the right ventricle measured 5 (11·25 mm. ·444 E. in.) and 5½ (11·85 mm. ·488 E. in.) French lines in thickness;—the average width in adults, of the two sexes, being 1·93 and 1·87 French lines.[1] The right auricle also undergoes great dilatation and its parietes become unusually thick. The left cavities on the contrary are relatively much smaller, and their walls thinner and more flaccid.

While, however, the chief valvular disease and the most marked hypertrophy and dilatation are found in the right cavities of the heart, the changes are not entirely limited to that side, but affect, to a less degree, the left also. Thus the folds of the mitral valve are not unfrequently found opaque and thickened, and the aortic valves may be in a similar condition, or the number of the curtains may be irregular. The hypertrophy and dilatation of the left ventricle are most marked in persons who survive for some years, and in such cases the general enlargement of the heart and the increase of weight which it attains are often considerable. In the two patients before referred to, the organ weighed 10 oz. and 17½ oz. avoirdupois respectively, though the normal weight, at the same ages, should not exceed eight or eight and a half ounces.

Some discussion has arisen as to the cause to which the hypertrophy of the right ventricle in cases of malformation is to be ascribed; and it has been supposed by Laennec, Bertin, Bouillaud, and others, to be due to the entrance of aërated blood through the aperture in the septum. This explanation cannot, however, be accepted as satisfactory.

---

[1] See Papers on the Weight and Dimensions of the Heart in Health and Disease, by the author, in the London and Edinburgh Monthly Journal of Medical Science for 1854.

When there exists any source of obstruction at the right
side of the heart, as in by far the majority of instances in
which the septum is imperfect, the course of the blood
through the aperture will be from the right ventricle into
the left; and it is probably only in the comparatively small
proportion of cases, in which, with a defective septum there
is no source of obstruction on the right side, that the reverse
obtains.   In the latter class of cases, however, the right
ventricle does not become hypertrophied to any great extent,
though the patient may survive for many years.   But
whenever there exists any source of obstruction at or
near the pulmonic orifice, and when the deficiency in the
septum is so great as to throw upon the right ventricle a
large share in the maintenance of the systemic circulation,
the right ventricle is invariably found more or less hypertrophied, even though the life of the patient may be prolonged only for a few weeks or months.   The ventricular
parietes also acquire as great an increase in width and firmness in the cases in which there is obstruction at or near
the pulmonic orifice without any defect in the inter-ventricular septum, and therefore without any entrance of
aërated blood, as when that partition is imperfect. It seems
therefore evident that the hypertrophy of the right ventricle
in these cases, is due to causes precisely similar to those
which occasion the same condition in after-life,—the increased growth which is consequent on the powerful muscular
efforts to overcome the obstruction occasioned by the contracted or rigid state of the pulmonic orifice, and to maintain
the circulation in the systemic as well as in the pulmonic
vessels.

It has been thought that in cases of malformation, the
right ventricle sometimes presents what has been termed
true " *concentric hypertrophy.*" It is now admitted, that, in
accordance with the observations of Cruveilhier and Dr.
Budd, in ordinary cases of hypertrophy the cavity affected
either retains its natural dimensions or undergoes dilatation,
and that increased thickness of the parietes with diminution
of capacity, is only observed when an unusually powerful

ventricle has been violently contracted at the period of death, as in some instances of rapidly fatal hæmorrhage.[1] The exception to this rule supposed to be afforded by cases of malformation, is also most probably only apparent and accidental. I have several times examined imperfectly developed hearts in which the right ventricle has been very small when the organ was first removed from the body, but, after they have been macerated till the muscular tension is relaxed, the cavity uniformly expands to a larger size and the thickness of the walls undergoes a proportionate diminution. Specimens exhibiting very thick parietes with small ventricular cavities, are, indeed, to be found in museums, but they have probably been placed in spirit before the muscular contraction had subsided, and thus the temporary state has been retained. The term concentric hypertrophy has also been applied to the condition of a ventricle from which the blood has been diverted into other channels, in consequence of which the cavity has become contracted and the parietes increased in width. The term is, however, still less applicable to this state, which is one of atrophy, not of hypertrophy.

Pathologists have also differed as to the relation in which in cases of this kind the deficiency in the septum of the ventricles, the origin of the aorta from the right ventricle, and the constriction of the pulmonary orifice, bear to each other. In the former edition of this work I adopted the explanation which seemed a necessary sequence to the views of Dr. Hunter as to the cause of the incompleteness of the septum, that the irregular origin of the aorta was the result of the obstruction to the exit of the blood from the right ventricle. Dr. Hunter pointed out that if during the earlier period of fœtal life when the septum of the ventricles was incomplete, some obstruction arose by which the right ventricle became incapable of freely transmitting the blood which it contained through the pulmonary artery into the

---

[1] See case by author, in Path. Trans., vol. i. 1846-47, 1847-48, p. 85.

ductus arteriosus and descending aorta, the current must
necessarily pass through the aperture in the septum into the
left ventricle, and thus the final separation of the two cavities
would be prevented.   To these views it is scarcely possible
to object, and it seems equally to result from them
that the septum in its further growth might be made to
deviate to the left, so that the aorta would communicate
with the right ventricle.   It also follows that from the connexion between the aorta and right ventricle, the constant
flow of blood from the cavity into the vessel must tend to
draw the aortic orifice still further to the right, so as to
produce the widening of the aperture and of the ascending
aorta which is so peculiar a feature in all those cases in
which that vessel has a double origin from the two ventricles.
A theory very similar to that here explained, and which has
been termed by Kussmaul the *overflow theory* (*Strauungs-
theorie*), was advanced by Dr. Mayer, of Zurich, almost
simultaneously with and quite independently of myself. It has,
however, recently been objected to by Drs. Carl Heine and
Halbertsma, who have supported a modification of the views
of Meckel.   The latter writer supposed that the incompleteness and displacement of the septum are primitive defects,
and that in consequence of the blood finding a ready
outlet from the right ventricle through the aorta, the pulmonary artery becomes more or less abortive, and the orifice
is contracted during the progress of development. This view
is supported by the fact pointed out by teratologists, and
which is well shown in the Icones Physiologicæ of Wagner
and Ecker, that, at the earlier period in the development of
the ovum the aorta is situated further to the right than
in the mature fœtus; though then, and even in the adult
heart, indications of the original position of the aorta
are retained.   The same theory affords an explanation of
the cases in which the aorta is found to arise wholly from
the right ventricle, while the pulmonary artery retains its
connexion with that cavity—a form of anomaly to which
the former does not equally apply.   It also explains a class
of cases to be hereafter mentioned, in which the pulmonary

artery, without any valvular defect, is abnormally small; but such cases are comparatively rare, and their production may be more satisfactorily accounted for in a different way. It however throws no light on the mode of formation of those very much more numerous cases in which, with the diminution in the capacity in the orifice and vessel, the valves of the pulmonary artery are seriously diseased. It leaves also the nature of the original defect unexplained, and substitutes for what appears in many cases an obvious and sufficient cause, an entirely theoretical idea. While therefore I am still disposed to regard the obstruction to the flow of blood from the right ventricle and the defect in the septum as standing in the relation of cause and effect; I think that the malposition of the aorta in these cases must be ascribed, partly to arrest of development, by which the earlier position of the vessel is retained, and partly to the obstruction to the flow of blood through the pulmonary artery and consequent distension of the right ventricle, causing the septum to deviate to the left. It is evident that provided the blood have direct access to the aorta from the right ventricle the origin of that vessel will be drawn further to the right, and the peculiar widening of the orifice will ensue. It is true that as pointed out by Kussmaul, the theory does not explain those cases of which several are on record, in which while the ventricular septum is deficient, the points of origin of the aorta and pulmonary artery are transposed. Such cases present, however, complicated anomalies, originating partly in arrest of development from obstruction at an early period of fœtal life, and partly in irregular evolution of the aorta and pulmonary artery from the primitive arterial trunk.

In the Museum of St. Thomas's Hospital there are various specimens exhibiting deficiency in the inter-ventricular septum, with obstruction at the pulmonic orifice. LL 67 is a preparation of this kind, removed from the body of a female child twelve months old, which had suffered from dyspnœa from birth, and had been livid and subject to frequent convulsive attacks. The pulmonary artery is of small size, and is provided with only two valves. The septum of

the ventricles is imperfect, so that the aorta is in communication with the right ventricle. The inter-auricular septum
is also very defective ; the left auricle imperfectly developed ;
and there are two superior venæ cavæ. LL 70 is a similar
specimen, except that the pulmonary valves are united
together, so as to form an infundibular or funnel-shaped
opening from the ventricle into the artery ; the ductus
arteriosus also is pervious and the foramen ovale open.
The preparation was probably obtained from a subject ten or
twelve years of age, but has no history attached to it.[1] In
LL 71 the pulmonary orifice is much contracted, owing to
the adhesion of the two valves with which it is provided, so
as to leave only a circular aperture. The septum ventriculorum is defective, and the aorta arises above the aperture
and communicates with both ventricles. The foramen ovale
and the ductus arteriosus are both closed. The preparation
was formerly in the Museum of Sir Astley Cooper. It was
removed from a boy, nine years and five months old, the
particulars of whose case, accompanied by a drawing of the
heart, are given by Dr. Farre.[2] At the time of birth
nothing unusual was observed in his appearance ; but, a few
months after, his complexion became dark ; and, at the age
of two and a half, his lips and cheeks had a bluish-black
colour, which was heightened by passions of the mind and
by cold. From this period till his death he was always
similarly affected, not only by mental causes but also by
slight corporeal exertion, particularly in very cold weather,
and his extremities were cool to the touch. Before he was
three years old he lost the use of the lower extremities, but
recovered under the care of Dr. Babington. He complained
of nausea and headache from the earliest period at which he
was able to express his feelings. Shortly before his death
he was seen by Sir Astley Cooper and Mr. Wheelwright.
He died of abscess in the right hemisphere of the brain with

[1] The form of the pulmonic valvular apparatus in this case is displayed in
plate 4, fig. 1.
[2] Malformations, p. 24. The case of John Cannon.

hemiplegia and convulsions. This case is further remarkable as displaying contraction at the commencement of the infundibular portion of the right ventricle and marked diminution of the cavity, so that it forms only a small passage immediately anterior to the orifice of the pulmonary artery. This condition will be more fully described hereafter.

In my own practice I have met with several instances of this description of malformation.

CASE I.[1]—*Great contraction of the orifice of the pulmonary artery ; aorta arising chiefly from the right ventricle; foramen ovale and ductus arteriosus closed.*

The boy who was the subject of this malformation first fell under my notice in June, 1846. He was then two years and one month old. I was informed that he was born healthy, and continued to thrive till he was vaccinated at the age of three months. Shortly after this he began to decline in health, and gradually became worse, till, when six months old, he was much in the same state as when brought to me.

His mother had previously had. two other children, both of whom were remarkably healthy. While pregnant with this child, and two months before her confinement, she was frightened by seeing a child killed, and never recovered the shock she sustained.

When first seen, the child's face was tumid, the cheeks of a deep rose colour, and the lips livid. The sternum was arched and prominent, and the ribs flattened above and expanded below. The abdomen was tumid, and the body generally emaciated. The hands and feet felt cold, and the fingers and toes were of a deep blue colour, and their extremities enlarged and club-shaped, more especially the

---

[1] Path. Trans., vol. i. p. 52 ; and Edin. Monthly Jour. of Med. Sc., vol. vii. 1847 (or N. S. vol. i.), p. 644. The preparation of the heart in this case is contained in the Museum of the Victoria Park Hospital, and is numbered B 4 in the Catalogue. It is represented in figs. 1 and 2, plate 2.

thumbs and great toes. The superficial veins in different parts of the body were very conspicuous. Several of the teeth were decayed, the mucous membrane of the mouth and tongue was in an unhealthy state, and the angles of the lips were ulcerated. There was also a livid excoriation around the anus. The pulse was extremely quick (136), and feeble and irritable. The cartilages of the ribs in the præcordia were prominent, and the dull space was increased in extent. A loud blowing and somewhat rough murmur, accompanying the impulse of the heart, was heard over the whole præcordia, and along the course of the sternum, on each side of the lower part of that bone, and in the epigastrium. It was perceived also in the neck and in the dorsal region on the left side of the spine. It was, however, most intense and rough in its character at the inner side of, and immediately above, the left nipple. The murmur was succeeded by a distinct second sound, but of a duller or flatter character than usual. The child was of very irritable disposition, and when unduly excited or fatigued by exertion, was subject to paroxysms of extreme difficulty of breathing, attended with violent palpitation of the heart; lividity, almost amounting to blackness, of the face, hands, and feet; and general turgescence of the superficial venous trunks.

For about two months the child was occasionally brought to me, and I afterwards lost sight of him till the 5th of October. I then found him much weaker than before; his head was large, and his mother thought had latterly much increased in size. He was greatly emaciated, and his appetite was extremely defective; his face pale; lips, hands, and feet livid; and the superficial veins, especially those of the neck, very large. The extremities were cold, and the fingers and toes more clubbed than before. The abdomen was tumid, and the mouth and anus still ulcerated. The præcordia yielded a dull sound from the third intercostal space to the edges of the ribs, and from the left of the sternum to the line of the nipple. The cartilages of the fourth, fifth, and sixth ribs were especially prominent. From the extreme irritability of the child, it was impossible accurately to investigate the physical signs; but the loud systolic murmur was heard very distinctly on the inner side of the nipple, and along the course of the sternum. The action of the heart was extremely rapid. On the evening of the day on which these notes were taken he was seized with convulsions, at first confined to the upper extremities, but subsequently becoming general. He continued sensible at intervals till the 7th, when he became comatose, and he died at two o'clock in the afternoon of the following day. During the last few hours he had violent palpitation of the heart, great lividity of the face and extremities, and extreme dyspnœa. He was two years and five months old at the time of his death.

The body was examined at noon on the 9th. The head was large and the anterior fontanelle somewhat open; much fluid was effused in the subarachnoid cellular tissue and into the ventricles. At the base of the brain a layer of soft and recently exuded lymph extended from the optic commissure to the posterior part of the pons Varolii, and passed for a considerable distance on each side, more especially in the course of the Sylvian vessels, so as to envelope the nerves proceeding from this part of the brain. In some places the deposit was fully two lines in depth. The subjacent cerebral substance, as also the parts contained in the ventricles, presented no appearances of disease. The brain weighed 37 oz. avoirdupois. The liver extended from the level of the fifth rib above, to an inch and a half below the edges of the cartilages, and across the entire upper part of the abdomen. It weighed 14 oz., and was of a deep purple colour, and very solid. The spleen was large and firm; the kidneys lobulated and congested; and the stomach and intestines healthy. Several small hæmorrhoidal excrescences were found around the anus, and the epithelium was there abraded.

The lungs were sparingly crepitant; the bronchial glands large, and the thymus also still of large size. The heart occupied its usual site; it was large for the age of the subject, and broader in its transverse than in its longitudinal axis. It measured in girth 5½ French inches (148·5 mm. 5·86 E. in.). The right ventricle occupied almost the entire front of the organ, and the great firmness of its walls presented a striking contrast to the softness and flaccidity of those of the left ventricle. The right cavities were distended with blood of a dark colour and tarry consistence. The right and left venæ innominatæ, the vena azygoso and the cavæ, were unusually capacious. The right auricle was much larger than the left, and its walls averaged nearly a line in thickness. The Eustachian valve was two or three lines in width, and the lining membrane of the auricle was opaque. The foramen ovale was entirely closed by its valve, and the fold formed a deep sac projecting into the left auricle. The right auriculo-ventricular aperture admitted a cylinder two inches and eleven lines in circumference (78·75 mm. 3·1 E. in.). The valves were thickened, but apparently competent to close the orifice. The sinus of the right ventricle was unusually capacious, and its walls were thick, measuring near the base 2½ French lines (5·6 mm. ·22 E. in.) at the midpoint 4 lines (9· mm. ·355 E. in.), and near the apex 2½ lines. The columnæ carneæ were large and firm. The infundibular extremity of the ventricle, though small relatively to the sinus, admitted a ball measuring one French inch in circumference (·27 mm. 1·06 E. in.); but the orifice of the pulmonary artery was contracted to a mere slit, two lines in length (4·5 mm. ·177 E. in.), situated between two valves, which were of a firm and fleshy character, and protruded forwards into the cavity of the artery. The trunk of the artery, though of very small calibre, was, in reference to the orifice, disproportionately large, and its coats were unusually thin. At the usual point it divided into three branches, two of which were distributed to the lungs; and the third—the ductus arteriosus—though at first pervious, became entirely obliterated towards its union with the coats of the aorta. A strong muscular band crossed the upper part of the ventricle, eight lines below the orign of the pulmonary artery, and in front of the auriculo-ventricular aperture, so as partially to divide the cavity into two portions; the anterior of which gave origin to the pulmonary artery; while that situated posteriorly opened directly and by a large orifice into the aorta.

The pulmonary veins entered the left auricle as usual. The left auriculo-ventricular aperture was less than the right, measuring only one inch and ten lines in circumference (49·5 mm. 1·95 E. in.); the valves were natural. The cavity of the left ventricle was also smaller, and its walls thinner than those of the right, measuring $2\frac{1}{2}$ (5·6 mm., ·22 E. in.), 3 (6·75 mm. ·26 E. in.), and 2 (4·5 mm. ·17 E. in.) lines in width. The columnæ carneæ were small and flaccid. The opening from the left ventricle into the aorta was indirect, and less than the communication between that vessel and the right ventricle. There could scarcely be said to be any deficiency in the inter-ventricular septum, though, through the aorta, the two cavities communicated. There were three semilunar valves at the orifice of the aorta, two of which corresponded with the right, and one with the left ventricle; the valves were natural. The ascending portion of the aorta was very large; it gave off the usual branches at the arch, and below the attachment of the impervious ductus arteriosus it diminished to about its natural calibre. The bronchial arteries were scarcely traceable, and the intercostal branches had their natural size.

This case affords a good example of the form of malformation in which the pulmonic orifice is obstructed and the aorta arises partly from the right ventricle. It differs, however, from the majority of such cases, in having had the foramen ovale closed and the septum of the auricles entire; and this, notwithstanding that the contraction of the pulmonic orifice was very marked.

The following case affords another instance of the same kind :—

E

CASE II.—*Contraction of the orifice of the pulmonary artery from fusion of the valves ; deficiency in the septum ventriculorum ; foramen ovale and ductus arteriosus closed.*

The boy, who was the subject of this malformation, first came under my notice in 1857, when he was four years old. His mother then stated that when born his lips were observed to be very blue, and he breathed rapidly. When six months old, his heart was noticed to beat unduly, and when he began to play about, he became subject on any active exertion or excitement, to violent attacks of dyspnœa and palpitation, in which he became livid in the face. These attacks occasionally terminated in convulsions, though when he became quiet they usually subsided. The veins of his head and chest were large, his fingers and toes clubbed, and the lips and extremities livid. A loud systolic murmur was audible about the left nipple, but not at the top of the sternum or at the posterior part of the left side of the chest.

He was frequently seen during the years 1857, 1858, and 1859, and upon the whole rather improved in health, being less livid and freer from the cardiac and convulsive symptoms. In the last year his friends went to reside in the country, and he was not under observation till February 1862. He was then found to be suffering more severely, apparently from having recently taken cold. He had a severe cough, and much difficulty of breathing, and was markedly cyanotic,—the conjunctivæ being injected, the cheeks flushed, and the lips and extremities livid ;—the fingers and toes were clubbed and the nails incurvated. A loud systolic murmur was heard over a large portion of the front of the chest, but was most distinct in the course of the pulmonary artery. He died convulsed shortly after being seen. He was then nine years old.

His body was examined the day after his death. The forehead was prominent, but the head not unduly large. The lower portion of the sternum was protruded, and the ribs depressed and flattened ; the extremities were very livid.

The lungs were dense and congestsd, but free from disease. The heart weighed four ounces avoirdupois. The orifice of the pulmonary artery was contracted from the adhesion of the valves, which were only two in number, so that it would only give passage to a cylinder fifteen French lines in circumference (33·75 mm., 1·33 E. in.). The deficiency in the number of the valves was evidently due to the blending of two of the segments. The right ventricle was of large size and its wall thick. The trunk of the pulmonary artery was small and its coats thin.

The septum of the ventricles was defective at its upper part, so that the aorta communicated with equal freedom with the two ventricles. The left ventricular cavity was small, and its walls thin and flaccid relatively to those of the right. The ascending aorta was of unusually large size. The foramen ovale and the ductus arteriosus were both completely closed.

The liver, spleen, and kidneys were large but healthy.[1]

The following case affords an instance of the more common variety of this malformation, in which the foramen ovale is open. For the particulars of the case I am indebted to Mr. Marshall, of Mitcham, in whose practice it occurred.

CASE III.[2]—*Deficiency in the septum ventriculorum; great contraction of the pulmonary orifice; open foramen ovale; ductus arteriosus closed.*

" G. S., aged six years and a half, was born at the full period; was rather dark coloured, and did not run alone till he was twenty months old. His mother stated that, at the fifth month of her pregnancy, she was much frightened.

[1] Path. Trans., vol. xiii. 1861–62, p. 57. Preparation in Victoria Park Hospital Museum, B 21.

[2] Path. Trans., vol. v. 1853–54, p. 67. The preparation of this case is preserved in the Museum of the Victoria Park Hospital, and is numbered B 6 in the Catalogue. The form of the pulmonic orifice is represented in plate 3, figs. 4 and 5.

When the child was three years and a half old, he had a fit,
and was paralysed on the right side.   The face soon
recovered ; but the arm and leg were ever after weak, particularly the former.   As the child has increased in size, so
have his symptoms become more severe.   His breathing has
become shorter, so as to be hurried on the slightest exertion,
and the surface has acquired a deeper tinge.   My first
acquaintance with the case occurred two years before the
boy's death, when he was brought to me suffering from
extreme dyspnœa, even when at rest, which was unusual
with him.   His condition at that time was as follows :—

" Skin of a purple hue, eyes large and appearing to start
from the orbits, conjunctivæ slightly injected, extremities
very purple, nails almost black, and ends of fingers much
flattened.   The chest was narrow and prominent in front.
The action of the heart was tumultuous and forcible, and
the sounds confused; the respiration was quickened.   A
few days before his death he took measles, and had passed
mildly through the disease, when he was seized with convulsions, occurring every few minutes and terminating in
coma, with dilated pupils and the eyelids half open.   The
fæces and urine were passed involuntarily, the surface of
the body was darker than usual, and the conjunctivæ
highly congested.   The boy died on the evening of July 31,
1853.   The body was inspected thirteen hours after death.
The head was not examined, but all the organs of the
chest and abdomen, except the heart, were found healthy,
though much congested."

The heart, which was sent to me by Mr. Marshall, was of
large size, and the right ventricle especially was hypertrophied and dilated.   The pulmonary orifice was so greatly
contracted owing to the adhesion of the valves, that only
a ball of six and a half French lines (14·6 mm. ·57 E. in.)
in circumference could be passed through it. The valves were
united into a hollow cylinder attached to the coats of the
vessel on one side, but elsewhere separated from them by a
deep sinus ; the passage from the ventricle into the vessel
being thus three or four lines long (6·75 to 9· mm.
·26 to ·35 E. in.).   The trunk of the artery also was of

small size. The septum of the ventricles was deficient at the base, over a large space, so that the aorta arose from both cavities. The opening from the right ventricle into the aorta had a capacity of twenty-one French lines (=7·25 mm. 1·86 ᴇ. in.); that from the left ventricle into the vessel, of twenty-four lines (54 mm. 2·13 ᴇ. in.). The foramen ovale was so freely open as to admit of the passage of the forefinger. The ductus arteriosus was closed.

In the following case the foramen was closed, but the duct was still pervious.

Case IV.[1]—*Contraction of the outlet of the right ventricle; imperfection of the septum ventriculorum; foramen ovale closed; ductus arteriosus pervious.*

R. B., a female, aged nineteen, was admitted into St. Thomas's Hospital in July, 1854. She was born at the full period, and from birth was of a peculiarly dark colour. When ten months old she began to suffer from violent fits of crying and excitement, in which she became black in the face; and these fits sometimes terminated in convulsions. The paroxysms recurred, at intervals of a few hours or days, till she was five years old when they ceased; but she continued subject to transient attacks of vertigo and headache. About the same time she had sores on the limbs for which she was sent to Brighton. After the cessation of the paroxysms she continued much in the same state till she was fourteen or fifteen years of age. She still suffered from palpitation and dyspnœa, aggravated by mental excitement or by corporeal exertion, and her health then became more impaired. When she was eighteen years of age, the catamenia appeared for the first time, and she had only three or four imperfect recurrences afterwards. She now became subject to frequent catarrhal attacks, and on several occasions spat large quantities of blood and bled freely from the nose.

---

[1] Path. Trans., vol. vii. 1855–56, p. 80. This preparation is numbered B 7 in the Catalogue of the Museum of the Victoria Park Hospital. The form of the pulmonic orifice is represented in plate 4, fig. 2.

Her mother stated that when she was pregnant with the child, she was greatly alarmed by her husband, who was insane, standing over her for two hours with a loaded pistol.

When admitted into St. Thomas's in July, she presented the usual symptoms of malformation of the heart. Her cheeks were livid, and the colour did not quickly return into them when they were blanched by compression. The conjunctivæ were injected; and the lips, gums, and interior of the mouth were of a dark purple hue, and the gums spongy and tender. The nails were much arched and dark-coloured, and the extremities of the fingers were somewhat bulbous. The chest was prominent in front, and flat at the sides. There was very obvious pulsation in the carotids, and the jugulars were distended but did not pulsate. She suffered from great difficulty of breathing and sense of suffocation, and was often incapable of lying down in bed. Her extremities were cold, and she felt constantly chilly. The pulse was quick (120), and the beat of the heart ill-pronounced. Her appetite and digestion were good, but the dyspnœa and palpitation were increased after taking food. The præcordial dulness commenced at the level of the third cartilage and continued to the edges of the ribs. Laterally it extended from the left side of the sternum to fully an inch beyond the left nipple. On listening over the third cartilage, between the nipple and sternum, a soft blowing murmur was heard very distinctly. It was audible, though less intensely, over the whole præcordia, and in a line towards the middle of the left clavicle. It was also distinctly heard to the right of the upper part of the sternum. A loud ringing second sound was heard about the middle and at the upper part of the sternum, after the murmur. A very marked thrill was felt on pressing the intercostal spaces over the whole præcordia, but especially between the third and fourth cartilages. The heart's sounds were heard in the left dorsal region, but the murmur was not there audible. The chest was sparingly resonant on both sides, and the respiratory sounds were harsh.

While in the hospital she did not improve, and she was discharged at her own desire. She was again admitted

after an absence of a month, at the end of August, when the
cholera was prevailing, and was then much worse; the lower
extremities, which were before somewhat swollen, were very
œdematous; the abdomen also was tumid, and the general
symptoms and physical signs were much as before.   Soon
after her re-admission she was seized with bilious vomiting,
the bowels being at the time confined.   She was relieved by
treatment, and after two days, when she had apparently recovered her ordinary state, she was suddenly attacked with
diarrhœa of the usual choleraic character, and rapidly sank
into fatal collapse.

On examination, the brain, lungs, liver, spleen, and
kidneys, were found greatly congested, but not otherwise
diseased.   The intestines contained the gruel-like fluid
which is commonly met with after death in the second stage
of cholera.   The heart was very greatly enlarged, weighing
17½ oz. avoirdupois.   The right auricle was distended with
uncoagulated blood; its cavity was greatly dilated and the
walls hypertrophied, measuring two to three French lines
(4·5 to 6·75 mm. ·17 to ·26 E. in.) in thickness.   The foramen ovale was closed, the fold being drawn up above the
isthmus and entirely adherent; the valve was somewhat
thickened.   The tricuspid valves were opaque and thick, and
studded with wartlike vegetations on their auricular side.
The right ventricle formed nearly the whole of the anterior
portion of the heart; its cavity was much enlarged, and its
walls firm and resistent and very thick, measuring from
three to seven lines (6·75 to 15·75 mm. ·26 E. in. to ·62
E. in.) in width in different parts.   It had two openings;
one of large size, from the sinus into the aorta; the other
very small, from the infundibular portion into the pulmonary
artery.   This aperture was so greatly constricted as only to
admit of the passage of a cylinder having a circumference
of eight French lines (18 mm. ·71 E. in.).   The constriction
was situated at the bases of the valves, and was formed by a
muscular band covered by fibrous tissue, and the edges of
the opening were studded with warty vegetations.   Immediately beyond the constriction the passage expanded, so
that the valves themselves freely admitted the forefinger

between them. The segments were two in number, and one
of them displayed some remains of a frenum or band on the
upper surface. Except being somewhat thickened and opaque,
they were free from disease. There was a deep sinus behind
each of them. The pulmonary artery was of small size.

The left auricle was small, and its walls averaged one line
and a half (3·75 mm. ·13 E. in.) in thickness. The mitral
aperture and valves were healthy. The left ventricle was less
capacious and its walls thinner and much less firm than those
of the right ; in width they ranged from two (4·5 mm. ·17
E. in.) to six (13·5 mm. ·53 E. in.) French lines. The aorta
arose much to the right of the pulmonary artery, and communicated with the sinus of the right ventricle by an aperture thirty-three French lines (75·37 mm. 2·93 E. in.) in
circumference, and with the base of the left ventricle by an
opening thirty-six French lines (81 mm. 3·19 E. in.) in circumference. The aortic valves were natural. The ascending
portion of the aorta was of large size, but the vessel diminished considerably after giving off the vessels to the head
and upper extremities. The ductus arteriosus was pervious,
and sufficiently large to give free passage to a crow-quill.

In 1849, I exhibited at the Pathological Society[1] the heart
of a child seventeen months old, which presented similar
malformations to those in the last cases. The child had been
under the care of Dr. Oldham, and had presented the usual
symptoms of morbus cæruleus. It died of effusion on the
brain preceded by icterus. The heart was very large for
the age of the child ; the pulmonary orifice was of small
capacity, and was provided with only two semilunar valves,
of which one displayed the appearances of imperfect division. There was an aperture at the base of the septum
ventriculorum, sufficiently large to allow of the passage of
the little finger, and the aorta arose above this aperture,
so as to communicate with both ventricles. The right
ventricle and auricle were very large, and the walls of
the ventricle were unusually strong and thick. The left

---

[1] Path. Trans., vol. ii. 1848–49, 1849–50, p. 37. The preparation is marked
B 11 in the Museum of the Victoria Park Hospital, and is represented in plate 1,
fig. 2.

cavities, on the contrary, were relatively small, and the parietes of the left ventricle were thinner and more yielding than those of the right. The foramen ovale was closed except at its upper and anterior part, where there existed a valvular opening, which, when the auricle was distended, would have allowed of the passage of blood from the right into the left auricle. The ductus arteriosus was pervious, but would only admit of the passage of a very fine probe.[1]

---

[1] In addition to the cases briefly noticed above, the following may be referred to. The precise seat and cause of the obstruction at the pulmonic orifice, and the state of the arterial duct, are often not mentioned in the reports.

Orifice or trunk of the Pulmonary Artery contracted ; Ventricular Septum imperfect ; Foramen Ovale open.

Ring, Med. and Phys. Jour., vol. xiii. 1805, p. 120. In a female child one year old.

Calliot and Duret, Bullet. de la Fac. de Méd. etc., 1807, No. 2, p. 21. In a boy aged eleven years.

Calliot, Bull. de la Fac. de Méd., etc., 1807, No. 2, p. 24. In a male child three years old. Duct. art. completely obliterated. Gintrac, obs. 21e. According to Obet (Bullet. des Sc. Méd., t. ii. Paris, 1809), the duct. art. was very small but pervious. The aorta made its turn across the right bronchus, and then passed behind the trachea to the left side of the spine. The duct. art. terminated in the left subclav. art. The latter vessel gave off the left carotid. The right carotid and subclavian arose separately. At the conclusion of the report by M. Obet is one by M. Huet, of Brest, of a case which corresponds with the case of Calliot and Duret, except as to the age, which is said to have been thirteen. This case is quoted as a distinct observation by Hein; De istis cordis deformationibus quæ sanguinem venosam cum arterioso misceri permittunt.—Gœttingæ, 1816, obs. 49e. It is, however, probably Calliot and Duret's case.

Palois, Bull. de la Fac. de Méd., etc., t. ii. 1809, No. 9, p. 133. In a male child aged four years. Gintrac, obs. 23e.

Knox, Ed. Med. and Surg. Jour., vol. xi. 1815, p. 57. In a cyanotic female child four years old. Duct. art. absent.

Ribes, Bullet de la Fac. de Méd., t. iv. 1815, p. 422. In a male child aged six years. Gintrac, obs. 37e.

Haase, Diss. Inaug. Med. de Morbo Cæruleo, Lipsiæ, 1813, p. 7. Hein, obs. 58. In a female who lived nine years and eleven months. Duct. art. nearly impervious.

Travers, Farre on Malformations, 1814, case of H. B., p. 34. In a male who lived four years, and was cyanotic and subject to dyspnœa on exertion. Pulmonary artery contracted to less than half its proper size. Duct. art. obliterated.

Leadam, Ibid., p. 37. In a male who lived twenty-two years, and was cyanotic and subject to difficulty of breathing on exertion. The pulmonary artery arose from the superior and central part of the right ventricle by a very narrow mouth of dense structure. Semilunar valves of pulmonary artery contracted by a warty excrescence, leaving passage in centre barely large enough to admit a small probe. Rest of pulmonary artery of proper size. Duct. art. closed.

Dorsay, New England Jour. of Med. and Surg., vol. i. 1812, p. 69. Girl, age at death not stated; symptoms commenced when two years old.

Gintrac, Sur la Cyanose, obs. 45ᵉ, p. 164. In a man twenty-one years of age, who had been cyanotic and died of phthisis. Pulmonary orifice constricted and had only two valves.

Cheevers, New England Jour. of Med. and Surg., vol. x. or N. S. vol. v. 1821, p. 217. In a male aged thirteen and a-half years. Duct. art. the size of a crow-quill.

Bloxham, Med. Gaz., vol. xv. 1835, p. 435. Iu a cyanotic female aged three years. Pulmonary artery very small and thin, and divided immediately after its origin.

Holst, Hufeland's Journal, 1837, quoted in Arch. Gén. de Méd., 2ᵐᵉ série, t. xi. 1836, p. 91. In a child aged two years. The duct. art. opened into the subclavian artery.

Tommasini, Clinica Medica di Bologna, quoted in Bouillaud, sur les Maladies du Cœur, 2ᵐᵉ éd. 1841, t. ii. p. 674, obs. 183. In a female aged twenty-five; only cyanosed at later period of life.

Lexis, Lancet, 1835-36, vol. ii. p. 433, with plate; quoted from a German journal. In a female child aged five years and three-quarters who died of phthisis. Duct. art. absent.

Napper, Lond. Med. Gaz., vol. xxvii. 1841, p. 793. In a boy aged five years aud seven months. Pulmonary art. admitted small probe.

Hildenbrand, Arch. Gén. de Méd, 3ᵐᵉ et nouv. série, t. xiv. 1842, p. 87. In a girl aged seven years. Left subclavian artery absent, and supplied by branches from the vertebral and left pulmonary arteries.

Landouzy, Ibid., 3ᵐᵉ série, t. iii. 1833, p. 436. Case of M. Magendie. In a cyanotic female aged eight years. Orifice of pulmonary artery 13 mm. iu circumference. Also Bullet. de la Soc. Anat., an. 13, 1838, p. 165.

Hope, Treatise on Diseases of the Heart, 3rd edition, 1839, p. 491. In a girl aged eight years. Pulmonary orifice contracted to size of a goose-quill.

Crampton, Trans. of the Dublin College of Physicians, N. S. vol. i. 1830, p. 34. In a boy, aged eleven years. Only a puckering of lining membrane in place of pulmonic valves.

Iliff, Dr. Chevers' Collection, p. 34. In a female, aged twelve years. Duct. art. closed.

Denucé, Bullet. de la Soc. Anat., an. 24, 1849, p. 124. In a cyanotic female, aged thirty-two months. Pulmonary orifice admitted small probe.

Valette, Gas. Méd. de Paris, 2ᵐᵉ serie, t. xiii. 1845, p. 97. Case under the care of M. Scoutettin. The preparation was shown to M. Sedillot, and is preserved in the museum at Strasburg. Female, lived five years. Pulmonary artery half the size of aorta.

Russell Reynolds, Path. Trans., vol. viii. 1856-57, p. 123. In a cyanotic female, aged thirteen months. Orifice of pulmonary artery very small, and only the rudiments of two valves. Two pulmonary veins entering the left auricle. Duct. art. closed.

Sanderson, Ibid., vol. x. 1858-59, p. 89. In a female child, aged six and a-half years. Cyanotic, and liable to attacks of dyspnœa and syncope. Heart weighed 3¼ oz.; pulmonary artery so contracted at its origin as barely to admit a small probe.

Nunneley, Ibid., vol. xiii. 1861-62, p. 42. In a female, cyanotic from birth, who died of phthisis when nearly fifteen years of age. Orifice of pulmonary artery a mere slit, not more than sufficient to admit a probe—vessel also small. Duct. art. closed.

Gubler, Comptes Rendus de la Soc. de Biologie, for 1861, p. 279, and Gaz. Méd. de Paris, 1862, p. 383. The patient died of phthisis.

Dufour, Bullet. de la Soc. Anat. de Paris, 27me année, 1852, p. 15. In a child subject to paroxysms of dyspnœa and cyanotic, aged eight years. Pulmonary artery nearly impermeable.

Mayer, Virchow, Arch. f. d. Path. Anat., 12ter Band, 1857, p. 497. In a child, aged eleven years and three-quarters. From early life cyanotic and liable to dyspnœa on exertion and to headache. Pulmonary artery remarkably narrow, only 14 mm. in diameter. Aorta passed over right bronchus and gave off art. innom. on left side and separate car. and subcl. arteries on the right. Duct. art. passing from left branch of pulm. art. to the left subclavian artery, but obliterated in its middle.

Orifice of the Pulmonary Artery contracted, &c.; Foramen Ovale not reported.

Farre, Malformations, 1814, p. 24. Case of Sir A. Cooper and Mr. Wheelwright. In a boy, aged nine and a-half years. Semilunar valves of pulmonary artery contracted into a small circle. Duct. art. closed.

Meyer, Hein, obs. 25, J. F. Meckel, Beitrag zur Geschichte der Bildungsfehler des Herzens welche die Bildung des rothen Blutes hindern.—Deutsches Archiv für die Physiologie.—Halle und Berlin, 1815, p. 221, obs. 42. In a female, aged seven years.

Gregory, Med. Chir. Trans., vol. xi. 1820, p. 296. In a cyanotic boy, aged eighteen years, who died of phthisis. Only slight contraction of the pulmonary artery.

Bertody and Dunglison, Phil. Med. Ex., 1845, quoted in Dublin Journal, vol. xxviii. 1845, p. 300. In a female, aged twenty-one years, who was cyanosed on exertion. Miliary tubercles in lungs. Pulmonary artery slightly contracted.

Jackson, American Jour. of Med. Sc., vol. xliii. 1849, p. 338. In a negro child, aged four years. Pulmonary artery small and only two valves.

Watson, Lectures, 2nd ed., 1845, vol. ii. p. 247. In a cyanotic male, aged seventeen years. Moderate contraction of the pulmonary orifices.

Maurin, Bull. de la Soc. Anat. de Paris, 28me année, 1853, p. 401. Pulmonary artery contracted. State of for. ov. and duct. art. not mentioned. In a boy, fourteen years of age, who had not been cyanotic.

Orifice of Pulmonary contracted, &c.; Foramen Ovale closed.

Olivry, Journal Gén. de Méd., t. lxxiii. 1820, 12me de la 2e série, p. 145. In a cyanotic male, aged six years. Gintrac, obs. 47e. Duct. art. closed. Pulmonary artery less than natural.

Louis, Arch. Gén. de Méd., 2me série, t. iii. 1823, obs. 9; and Mem. et Rech. Anatomico-Pathologiques, 1826, p. 313, obs. 10. In a man, aged twenty-five years. Pulmonary orifice greatly contracted by a diaphragm ; combined with contraction of right auriculo-ventricular aperture.

Graves and Houston, Dublin Hospital Reports, vol. v. 1830, p. 322, Case 1. In a boy, aged three years. Pulmonary artery half its natural size. Duct. art. pervious, but narrow.

Blackmore, Ed. Med. and Surg. Jour., vol. xxxiii. 1830, p. 268. In a girl, aged three and a-half years. Orifice of pulmonary artery small. Great contraction, or possibly obliteration of the left auriculo-ventricular aperture. No duct. art.

Marshall, Lond. Med. Gaz., vol. vi. 1830, p. 886. In a man, aged twenty-three years.

Huss, Gaz. Méd. de Paris, t. xi. 1843, p. 91. In a cyanotic boy, aged six years. No appearance of arterial duct. Pulmonary orifice two lines in diameter. Two valves. Tubercles in lungs.

Gravina, Schmidt's Jahrbücher, 1839, quoted in Arch. Gén. de Méd., 3me et nouv. série, t. vi. 1839, p. 360, and Ed. Med. and Surg. J., vol. liii. 1840, p. 522. In a boy, aged nine years.

Chevers, Collection of Facts, etc., p. 36. In a boy, aged sixteen years.

Chevers, Ibid., p. 36. In a person aged fifteen or sixteen years. Possibly with an open ductus arteriosus.

Escalier, Bullet. de la Soc. Anat., 20me an. 1845, p. 213. In a cyanotic female, aged eleven years.

Dalrymple, Path. Trans., vol. i., 1846-47, 1847-48, p. 58. In a female, aged twenty-five years.

Hare, Path. Trans., vol. xi. 1859-60, p. 45. In a slightly cyanotic male infant, which lived two months, pulmonary artery very small and orifice only three-eighths of an inch in circumference, and provided with three rudimentary valves. Foramen ovale closed, except for an orifice as small as a pin point.

Bouillaud, Bullet. de l'Acad. Imp. de Méd., t. xxviii. 1862-63, p. 777. Contraction of the orifice of the pulmonary artery, septum of the ventricles entirely wanting, but the auricular cavities distinct. In a man, aged thirty-nine years, who was not cyanotic. This case, from the very imperfect state of the partition of the ventricles, would appear to be closely allied to those treated of at an earlier period of the work, p. 16, et infra.

Dubreuil, Des Anomalies artérielles, plate 1, fig. 1, in a cyanotic female, aged nine years, who died at the Bordeaux Hospital. The specimen is preserved in the museum. Orifice of pulmonary artery nearly obliterated. No appearance of duct. See also Föster, Taf. xix., figs. 15 and 16.

2. *Obliteration or Atresia[1] of the orifice and trunk of the pulmonary artery.*

A more aggravated degree of the kind of malformation to which allusion has just been made, is that in which the orifice or trunk of the pulmonary artery is entirely impervious. A case of this description was described by Dr. Hunter, in 1783, in the paper before referred to.[2] The child was born at the eighth month, was very livid, had violent palpitation, and died in convulsions on the thirteenth day.

---

[1] ά τρησις, a piercing or boring through.
[2] Med. Obs. and Enq., vol. vi. 1783, p. 291, case 1.

The pulmonary artery was found entirely impervious, and contracted into a solid cord. The septum of the ventricles was entire; and the right ventricle had scarcely any cavity remaining, while the left was large and powerful. The foramen ovale continued open, and the pulmonary branches received their supply of blood from the aorta through the medium of the arterial duct. In 1812, a specimen, similar so far as the obliteration of the pulmonary orifice is concerned, was briefly described in the *London Medical Review;* and in 1814 some further particulars of the same case, of which the preparation was in the possession of Mr. Hodgson, were given by Dr. Farre.[1] The pulmonary artery was reduced to an impervious filament leading to the ductus arteriosus. The latter vessel, which was of large size, was connected with the aorta, and gave off the pulmonary branches. In the septum of the ventricles " some of the muscular fibres were wanting, and the lining membrane of the left ventricle had three foramina, giving it a cribriform appearance." The foramen ovale was largely open, and the right auriculo-ventricular aperture and the corresponding ventricle were of small size. The left ventricle, on the contrary, was unusually large. The child was observed to be of a deep purple colour soon after birth, and had difficulty of breathing. It died in convulsions on the seventh day. In the same publication Dr. Farre[2] alluded to two other cases which had occurred to Mr. Langstaff. One of these was that of a still-born child. In the other, the infant lived six months, and its temperature was always much below the natural standard; its skin of a deep colour, almost black; and it had fits daily. In this instance the septum of the ventricle presented a considerable perforation; but the opening, and the cavity of the right ventricle generally, was almost filled up by muscular fibres. The ductus arteriosus was remarkably short and very small. Dr. Farre mentions that he had examined a third specimen

---

[1] London Med. Rev., vol. v. 1812, p. 262; and Farre on Malformations, p. 19.
[2] On Malformations, etc., pp. 19 and 27.

exhibiting this defect, and that a fourth had been reported
to him.   In 1816,[1] Mr. Howship published a case which
agreed with that of Mr. Langstaff in the circumstance of
the septum cordis being deficient; but the aorta arose above
the aperture, and the right ventricle was unusually large
and powerful, while the left cavities were of small size. The
child, which first presented symptoms of cyanosis when fifteen days old, lived to the age of six months.

Since the publication of these cases, others of a similar
description have been recorded both in this country and
elsewhere; and specimens have been exhibited at the Pathological Society, by Drs. Crisp,[2] Chevers,[3] and Hare;[4] and I
shall have again occasion to refer to two other cases shown
there by myself.

Since the publication of the last edition of this work, an
interesting case of atresia of the orifice has been published
by Dr. C. Heine, of Tübingen. The child which was the
subject of the malformation was born at the full period, but
was deeply cyanotic and survived for only two days. The
orifice of the pulmonary artery was entirely closed by a
membranous septum, in which there were no traces of segments. The aorta arose from the right ventricle, and communicated with the left through an aperture in the inter-
ventricular septum. The foramen ovale was open, and the
ductus arteriosus furnished the supply of blood to the lungs.[5]

---

[1] Practical Observations in Surgery and Morbid Anatomy, 1816, p. 93.

[2] Path. Trans., vol. i. 1846-47, 1847-48, p. 50.   In a female twelve years of
age, cyanosed and subject to cardiac symptoms from birth.   The pulmonary
artery was entirely absent.   There were two small vessels which terminated
in blind extremities on the parietes of the heart (Cor. Art.)   The source from
which the pulmonary supply was derived is not named.   See also Structure of
the Bloodvessels, p. 92.

[3] Ibid., p. 204.   Orifice of the pulmonary artery obliterated by fusion of the
valves.   The trunk was small and connected with the aorta by a vessel—the
ductus arteriosus.   From an infant.   Dr. Chevers thinks there was probably
some other supply of blood to the lungs.

[4] Ibid., vol. iv. 1852-53, p. 81.

[5] Angeborne Atresie des Ostium arteriosum dextrum, Tübingen, 1861, p. 23,
where reference is made to another case of somewhat similar anomaly.

A peculiarly interesting example of this description of malformation has also been recently placed on record by M. Raoul-Chassinat. It occurred in a child which only lived twelve days, and had a hernia of the liver, but did not present any symptoms of defect in the circulating organs. The heart consisted of three ventricles, or more properly the right ventricle was divided into two cavities in the way to be hereafter described; one portion of which communicated with the left ventricle and so indirectly with the aorta, while the orifice and a part of the trunk of the pulmonary artery was obliterated. The left auricle received only one pulmonary vein, while another passed through the diaphragm and entered the vena cava ascendens.[1]

It will be seen from the cases which have been briefly alluded to, that the obliteration is sometimes confined to the orifice of the pulmonary artery. In such instances it is caused by the union of the valves, forming a kind of septum or diaphragm stretched across the opening, and entirely separating the cavity of the vessel from that of the ventricle; while the trunk of the artery remains pervious. In other cases a larger or smaller portion of the pulmonary artery, commencing at its origin from the ventricle and extending towards the ductus arteriosus, is impervious and converted into a ligamentous cord. In some of these cases the obliteration is probably due to inflammation of the lining membrane of the vessel. In a third class of cases the obliteration would appear to be originally seated in the outlet of the ventricle, and to be caused by disease of the endocardium or fibrous zone in that situation, or by hypertrophy of the muscular substance.

Obliteration of the pulmonary orifice, it will further be seen, may occur at different periods of fœtal life; either in the early months before the septum cordis is completely formed, or during the later periods after the separation of the ventricles is completed. It is, however, of much the most frequent occurrence at the former period. Indeed, of

---

[1] Arch. Gén. de Méd., 2me série, t. xi. 1836, p. 80.

thirty-four cases of this anomaly of which I have collected
notes, in eight only, including that of Dr. Hunter, does the
disease appear to have occurred when the septum of the
ventricles was already completed ; though a ninth case of
the same kind is referred to by Dr. Chevers, as existing in
the collection at Guy's Hospital. In most of the cases cited
the defect in the septum of the ventricles appears to have
been considerable ; but the preparation in the possession of
Mr. Hodgson formed an exception to this rule.

If the obliteration occur while the growth of the septum
is in progress, the right ventricle becomes large and powerful, and the aorta derives its chief supply of blood from that
cavity ; while the left auricle and ventricle are disproportionately small. If, on the contrary, the obliteration occur
after the septum of the ventricles is completed, the right
ventricle and right auriculo-ventricular aperture become
very much diminished in size, and the left cavities are unusually large and powerful. The latter condition also
obtains when, as in the case of Mr. Hodgson, the communication between the ventricles is not free.

When the septum of the ventricles is imperfect, the
foramen ovale is occasionally found closed. Thus there
was no communication between the auricles in four out of
the twenty cases referred to, and in two others the state
of the foramen is not named in the reports. When, on the
other hand, the septum of the ventricles is entire, the foramen ovale is necessarily pervious to some extent ; though,
as in the case exhibited at the Pathological Society by Dr.
Hare, the opening may only be of small size.

Generally in cases of obliteration of the pulmonary artery,
the blood is transmitted to the lungs from the aorta through
the ductus arteriosus. Indeed, in twenty out of twenty-eight
cases, this seems to have been the channel through which
the supply was obtained, though such is not in every instance expressly stated to have been the case. In one
instance, that related by M. Lediberder, the duct did not
exist in its usual situation ; but there was an aperture in
the trunk of the pulmonary artery, immediately below its

bifurcation, by which it communicated directly with the aorta, and so the pulmonary branches obtained their supply. The opening was in this case also doubtless the analogue of the ductus arteriosus, though the vessel itself had become abortive. The case, indeed, bears a close analogy to those in which the pulmonary branches are given off directly from the aorta. An instance of this kind which I have had the opportunity of examining at the Children's Hospital, was exhibited by Dr. Buchanan at the Pathological Society, in 1864.[1]

In two cases the ductus arteriosus is reported to have been closed, and the sources of the pulmonary supply were not ascertained. In a third case, in which the duct was closed—that of Dr. T. K. Chambers, as described by Dr. Chevers[2]—the pulmonary artery is supposed to have received its blood from the left subclavian artery; and in that of Dr. Babington[3]—also related by Dr. Chevers—the supply furnished to the lungs through the ductus arteriosus was supplemented by other vessels arising from the aorta. The blood circulating through the duct was distributed by two branches to the right lung, and by one very small and long vessel to the left lung. The left branch of the pulmonary artery was obstructed; and the left lung received its supply partly through the small vessel, and partly through one of two additional branches which arose from the descending aorta. The other branch passed to the right lung. The period during which the patients survived in these instances was remarkable, both having lived to the age of nine or ten. In a case related by Dr. Shearman,[4] however, in which the pulmonary artery was not obliterated, but was supposed not to have transmitted any blood, and the source of the pulmonary supply

---

[1] Path. Trans., vol. xv. 1863–64, p. 89. From a child which lived three months. There was no lividity, but the child suffered from cardiac symptoms. The foramen ovale was largely open, and the septum of the ventricles incomplete.
[2] Collection of Facts, p. 15.
[3] Ibid., p. 14. The preparation is marked 1383^{25} in Guy's Hospital Museum.
[4] Prov. Med. and Surg. Journal, 1845, p. 484.

was not ascertained, the patient lived to the age of nine. Dr. Crisp has also related a case in which the ductus arteriosus most probably conveyed the blood to the lungs, and the girl attained the age of twelve years. In all these cases, however, communication could freely take place between the two sides of the heart through the opening in the septum of the ventricles, and in all but one the foramen ovale was also unclosed.

A case, which is described as one of obliterated pulmonary artery is reported in the sixty-first volume of the Medical and Physical Journal,[1] and a preparation, which is probably that of the heart described, exists in the Museum of the London Hospital. By the kindness of the late Mr. Nathaniel Ward, I had the opportunity of examining this specimen, and found that it presented a condition, so far as I am aware, unique. The pulmonary artery is not obliterated, but exists as a vessel of very small size throughout its whole course. The aperture of communication with the ventricle is a mere slit in an opaque membrane stretched across the orifice, which will only admit the end of a small probe. This membrane is formed by the adhesion of the valves, which are apparently only two in number. The artery itself will admit a large crow-quill; and it divides into the usual branches distributed to the lungs. There is no trace of the ductus arteriosus; but, about an inch below the origin of the left subclavian artery, a vessel arises from the aorta, the size of a goose-quill, which divides into two branches distributed to the left lung. About half an inch lower down a second but smaller vessel is given off, which is distributed to the right lung. The aorta is of large size, and is placed above a considerable aperture in the septum of the ventricles, so that it freely communicates with both cavities. The left coronary artery arises at a higher point than natural. The ventricles are both of large size. The foramen ovale is closed, but the Eustachian valve is fully developed. The specimen is stated in the catalogue to have

---

[1] P. 548.

been removed from the body of a female, sixteen years of
age, who died of phthisis in the practice of Dr. Ramsbotham,
and is doubtless that referred to in the Journal. The patient
had during life presented the usual features of morbus
cœruleus.   In this instance, it is probable that the original
defect consisted in the premature obliteration of the ductus
arteriosus, and a case exhibited by Dr. Quain for Dr. Sibbald
at the Pathological Society, and described in the Transactions,[1] and that of Dr. Buchanan before referred to, may
have had a similar origin.

Cases of this description depend on the faulty development
of the branchial arches, by which the portions which should
form the ductus arteriosus and pulmonary artery become
obliterated, to a greater or less extent, at an early period of
fœtal life.   In all cases in which the pulmonary artery is
completely obstructed, while the ductus arteriosus remains
pervious, that vessel must necessarily become the channel
through which the blood is conveyed to the lungs; though
in some such cases the pulmonary branches may appear
to be derived from the aorta by an independent vessel.
If, however, the portion of the branchial arch which
forms the ductus arteriosus becomes prematurely obliterated, the pulmonary artery, conveying during fœtal
life only the blood transmitted to the lungs, will remain
very small, and may require other vessels to furnish a
compensatory supply; or, if the obliteration involve both
the ductus arteriosus and the pulmonary artery, the supply
to the lungs may be entirely derived from other sources. In
the latter class of cases, the vicarious supply will probably
be furnished by the bronchial arteries, for those vessels
naturally anastomose with the pulmonary artery, and at an
early period of fœtal life they may be readily capable of
undergoing the requisite dilatation.

In cases in which the blood is transmitted to the pulmonary branches through the ductus arteriosus, that vessel

[1] Vol. viii. 1856–57, p. 167.   This patient died at the age of ten months, and
the septum of the ventricles was imperfect.

is sometimes very short; and, indeed, may be so reduced in size as only to appear in the form of an opening leading directly into the pulmonary artery from the aorta, as in the case of M. Lediberder. A still further degree of atrophy may be supposed to occasion these cases in which there is no remains of the ductus arteriosus and the pulmonary branches arise as distinct vessels from the aorta.[1]

The following cases of obliteration of the pulmonary artery have fallen under my own notice.

CASE V.—*Obliteration of the orifice and trunk of the pulmonary artery; aorta arising chiefly from the right ventricle and giving off the pulmonary branches through the ductus arteriosus.*[2]

The subject of this case was a male child under the care of Dr. Bentley, at the City Dispensary, in 1847, which I had several times an opportunity of seeing and examining. His mother stated that, at the time of birth, his skin was extremely dark-coloured, but he breathed freely. The child was from the first, restless and peevish; and, when three months old, began to suffer occasionally from paroxysms of extreme excitement and crying, during which his breathing became very rapid and laborious; there was violent palpitation of the heart and of the vessels of the neck and scalp, and great lividity of the face, with convulsive contractions of the hands and feet.

When first seen by myself on November 15, 1847, the

[1] Dr. Buchanan's case, Path. Trans., vol. xv. p. 89, and page 65 supra.

This case is described as an example of the primitive condition of the arterial vessel being retained, but as the coronary arteries are derived from the vessel in their natural situation, it is evident that the existing trunk is the aorta, the pulmonary artery and ductus arteriosus having doubtless become abortive at an early period of fœtal life.

[2] Path. Trans., vol. i. 1846-47, 1847-48, p. 205. The preparation is contained in the Victoria Park Hospital Museum, and is marked B 8. It is also delineated in plate 5, fig. 1.

child was ten months old. He was then tolerably well nourished, and exhibited sufficiently characteristic, though not very marked, cyanosis. The lips were of a purplish colour, the lower eyelids dark, and the face pale and puffy. The extremities of the fingers and toes were somewhat bulbous, and the glandular portions of the finger and toe nails nearly black. The sternum was arched and prominent, and the lower ribs were much drawn in with the inspiratory act. There was some fulness, but no pulsation, of the jugular veins. A loud systolic murmur was audible over a large portion of the front of the chest, and less distinctly posteriorly on the left side of the spine in the interscapular space ; but the child was in so excited a state —crying continually and incapable of being quieted for a moment—that it was impossible accurately to investigate the physical signs. The head was of very unnatural form, being remarkably narrow and contracted anteriorly, and broad and elongated behind. The anterior fontanelle was closed, but the skull was very prominent in that situation.

The child continued to suffer from the attacks of dyspnœa which have been described, generally once daily, sometimes more frequently, according to the degree of severity and length of duration ; a long interval existing between the paroxysms only when it had been completely exhausted. Towards the end of the seizures the breathing became quick and laborious with a tendency to suffocation, and the face and extremities were extremely livid. The tongue was drawn up to the roof of the mouth, and the thumbs were pressed into the palm of the hand. The child refused the breast ; and it died in one of these attacks, on the 11th of January, 1848, when wanting two weeks to being twelve months old. The father was a healthy man ; but the mother was much out of health, and stated that when pregnant with this child, she was greatly alarmed by seeing a man who was dying of asthma. They had but one other child, and that, though very healthy looking, was very backward for its age—between three and four years—being still incapable of speaking.

On the examination of the body, the integuments were
found sufficiently provided with fat. The lungs were small,
very much congested, and sparingly crepitant.

The pericardium was natural; the heart was of very
large size for the subject, being much expanded transversely,
and weighing 3¼ ounces avoirdupois. The right auricle
was extremely large, and its walls unusually thick. The
foramen ovale was covered by its valve, which, however, not
being adherent to the isthmus, would allow the flow of
blood from the distended right auricle into the left. The
Eustachian valve was very imperfectly developed. The
right auriculo-ventricular aperture and its walls were natural.
The cavity of the right ventricle was of very large size, and
consisted almost entirely of the sinus; the infundibular
portion was reduced to a mere chink and was entirely closed
at the usual point of origin of the pulmonary artery. The
trunk of the pulmonary artery formed an impervious cord,
extending from the ventricle to the bifurcation of the vessel
and its union with the ductus arteriosus. The septum of
the ventricle was imperfect at the base. The walls of the
right ventricle were extremely thick, measuring five to six
Paris lines (11·25 to 13·5 mm. ·44 to ·53 e. in.) in width,
and unusually firm. Relatively to the right cavities, the
left auricle and ventricle were very small, and the walls of
the latter were thinner and more flaccid. The valves were
natural. The aorta arose in chief part from the right ventricle, and was of large capacity so far as the point at which
it gave off a vessel, evidently the ductus arteriosus, through
which the supply of blood had been transmitted to the
lungs. This vessel communicated with the aorta at a point
somewhat anterior to the natural situation of the duct, and
was of sufficient size to admit a small goose-quill. After
following a course of six lines, it divided into two branches
transmitted to the right and left lungs, and from the point
of bifurcation a small but solid cord, the impervious trunk
of the pulmonary artery, passed down to the base of the
right ventricle in the usual situation of the pulmonary
artery. The coronary arteries and pulmonary veins were

natural. The abdominal viscera were much engorged, and the parenchymatous organs enlarged and solid.

The case which has been related affords an example of obliteration of the pulmonary orifice occurring at the earlier period of fœtal life, before the septum of the ventricles is completed. Of the rarer description of cases in which the obliteration occurs after the division of the ventricles is effected, a specimen was exhibited at the Pathological Society by Dr. Hare in 1853.[1] It was removed from a male child, who was very livid, and died when nine months old. The right auricle was of large size, and the foramen ovale was open, but only to the extent of one-sixteenth of an inch in breadth, and one-tenth or one-twelfth of an inch in length. " On cutting into the right ventricle, it was found that the columnæ carneæ were fused almost into one," so that it presented nearly " a solid mass." The cavity would only hold a moderate-sized pea. The septum ventriculorum was perfect. The orifice of the pulmonary artery was entirely closed, but its trunk constituted a cul-de-sac in communication with the ductus arteriosus, and divided into the usual branches. The opening of the latter passage from the aorta would admit a crow-quill. The left ventricular walls were very thick, measuring fully half an inch, and that cavity gave as usual origin to the aorta.

The most interesting feature in the case was the extremely small size of the opening in the foramen ovale, considering that the passage afforded the only medium of communication between the right and left sides of the heart.

The following case also affords an instance of the form of defect arising after the completion of the septum, and involving only the orifice of the pulmonary artery :—

---

[1] This specimen is preserved in the Museum of University College, London.

CASE VI.—*Obliteration of the orifice of the pulmonic artery from adhesion of the valves; septum of the ventricles entire, foramen ovale and ductus arteriosus pervious.*[1]

For the opportunity of examining this specimen I was indebted to Dr. Lanchester, the late resident medical officer of the Victoria Park Hospital for Diseases of the Chest. The infant from which the heart was removed, was taken to Dr. Saul, of the St. Pancras Workhouse, to be vaccinated, and was observed to be cyanotic. It died when only nine days old.

The heart was of unusual form, being broader from side to side than from above downwards. In the former direction it measured one inch and nine lines French (47·25 mm., 1·79 E. in.); in the latter one inch and four lines (36·1 mm., 1·35 E. in.). The left ventricle constituted much the largest part of the organ. The muscular substance of the right ventricle was soft and yielding, that of the left was firm and resistent. The two auricles communicated freely through the unclosed foramen ovale. The cavity of the right ventricle was of very small size. The outlet from the ventricle by the pulmonary artery was entirely closed by the union of the valves at the origin of that vessel. The pulmonary artery was pervious down to the valves, which were entirely adherent together and showed on the arterial side three ridges with corresponding depressions between them, indicating the forms of the original valves and the lines of attachment. The ductus arteriosus of the usual size, passed into the aorta, and formed a communication between the branches of the pulmonary artery and that vessel. The septum of the ventricles was entire. The cavity of the left ventricle was of large size and was separated from the left auricle by the usual valves. The ascending aorta was large and the ordinary branches arose at the arch. After the entrance of the ductus arteriosus the aorta diminished considerably in capacity.

---

[1] From Path. Trans., vol. xv. 1863-64, p. 60. Preparation in Victoria Park Museum, B 25.

The course of the blood must in this case have been from the right auricle into the left auricle, thence into the left ventricle and aorta, and from that vessel to the lungs by the ductus arteriosus.   The right ventricle, though communicating naturally with the right auricle, yet having had no passage through it, had become atrophied ; while the left, having maintained both the systemic and pulmonic circulations, was unusually large and its walls thick and firm. Only a small portion of the blood circulating in the body could have been subjected to the influence of the air in the lungs.

The heart after being macerated in water weighed 16¾ drachms avoirdupois.   The lungs were fully expanded.   The other organs were not examined.

It has been supposed by Dr. Heine that the obliteration of the orifice of the pulmonary artery by the adhesion of the valves, as occurred in his case and the one here recorded, is of less frequent occurrence than the obstruction of a larger or smaller portion of the trunk of the vessel.   It does not, however, appear that the rarity of these cases is so great as is supposed by Dr. Heine ; indeed, ten or eleven of the cases of obstruction appear to have been instances in which the aperture only was obliterated.   It seems also highly probable that when the trunk of the vessel is involved, the disease generally commences in the orifice and gradually extends towards the other extremity.   It appears to be unusual for the orifice to be free and the trunk of the vessel obstructed. The particular situation occupied by the obstruction is, however, of much less importance in cases of this description than the state of the inter-ventricular septum and the condition of the fœtal passages,—the foramen ovale and the ductus arteriosus,—as upon the persistence of one or other of these passages depends the possibility of the circulation being at all maintained.   In by far the largest number of recorded cases—all, indeed, but eight or nine—the septum of the ventricles is reported to have been imperfect, so that the aorta derived its blood from both ventricles; thus indicating that the defect at the pulmonic orifice occurred at the early period of fœtal life. When this is the case the blood readily passes from the right to the left side of the heart, and so enters the systemic circulation. In the comparatively few cases in which the obliteration occurs after the completion of the septum, the foramen ovale must necessarily form the channel of communication between the two sides of the heart. By this means, however, the circulation is apparently less readily maintained than in the former case; the children in whom such a condition existed having all died very early, while some of those in whom the septum of the ventricles was imperfect survived for several years.[1]

---

[1] In addition to the cases of *obliteration or atresia of the orifice or trunk of the pulmonary artery* quoted above, the following may be referred to :—

With the Septum of the Ventricles entire.

M. Lordat, Gintrac, obs. 53°. The child lived six weeks. The pulmonary artery very small and obliterated at its orifice, arterial canal natural.

Dr. Carson, Edin. Med. and Surg. Jour., vol. lxii. 1844, p. 134; and Dublin Jour., vol. xxvi. 1845, p. 126. In a child that lived five days. Pulmonary artery wanting, pulmonary arteries supplied from aorta.

Ollivier, Bullct. de la Soc. Anat., vol. xxxvi. année 1861, N. S., t. vi. p. 320. In a child which lived two days. Orifice of pulmonary artery obliterated, duct. art. and for. ov. open, right ventricle almost obliterated.

Gueniot, Ibid., t. xxxvii. année 1861, N. S., t. vii. In a child which lived fourteen days. Pulmonary orifice obliterated, but branches pervious. For. ov. and duct. art. open. Right ventricle almost abortive.

Hall and Vrolik, quoted in Arch. Gén. de Méd., t. viii. 1825, p. 595. In a female child which lived two years. The pulmonary artery obliterated at its origin from the ventricle, but a second vessel passing from the ventricle to the aorta. Duct. art. open.

With the Septum of the Ventricles imperfect.

Houston, Dublin Hospital Reports, vol. v. 1830, p. 324, case 2nd. In a child which lived eighteen months. The pulmonary artery was small but pervious to near its origin, where it was obliterated. Duct. art. and for. ov. open.

Fearn, Lancet, 1835, vol. i. p. 312. In a child which lived seven weeks. Orifice of the pulmonary artery closed by a thin membrane. The ductus arteriosus is stated to have been closed, and the means by which the lungs were supplied with blood was not ascertained.

Spital, Edin. Med. and Surg. Jour., vol. xlix. p. 109. In a child which lived twenty-three days. For. ov. and duct. art. open.

Biggar, quoted in Med. and Surg. Jour., vol. lv. p. 251, from German publication. The child lived five and a half months. Vessel from arch of aorta which went to lungs? Duct. art.

J. G. Smith, Lancet, vol. i. 1841–42, p. 543. In a child which survived eight months. A small rudimentary pulmonary artery passing to the ductus arteriosus.

3. *Constriction at the commencement of the infundibular portion of the right ventricle.*

Writers on cardiac malformations refer to the occasional occurrence of supernumerary cavities in the heart, and in some of the cases described such excess may have existed. Thus Kerkring[1] states that he has seen the right ventricle double with two pulmonary arteries. Probably, however, by far the majority of instances of such anomaly consist in the development of septa in the cavities. Indeed, apparent

---

Douglas, Med. Gaz., vol. xxxi. 1848, p. 16. The boy lived eighteen months. Duct. art. and for. ov. largely open.

Dr. Chevers, Path. Trans., vol. i. 1846–47, 1847–48, p. 204. In a child apparently several weeks old, but without any history. Duct. art. very small.

Dr. Crisp, Path. Trans., vol. i. 1846–47, 1847–48, p. 50 ; and Diseases of Bloodvessels, p. 92. In a female twelve years old. Duct. not named. For. ov. closed.

Lediberder, Bullet. de la Soc. Anat., 11me année, 1836, p. 68. In a boy twelve days old. The ductus arteriosus did not exist in its usual situation ; but a direct opening was found between the aorta and pulmonary artery.

Laurence, Ibid., 12me année, 1837, p. 216. In a female child which lived fifteen days. Pulmonary artery obliterated at its origin. Duct. supplied pulmonary vessels.

Shearman, Prov. Med. and Surg. Jour., 1845, p. 484. Pulmonary artery very small and valves rudimentary. The vessel had probably not given passage to any blood. In a girl nine years of age. In this case the ductus arteriosus is stated to have been closed ; but the source of pulmonary supply was not ascertained.

Graily Hewitt, Path. Trans., vol. viii., 1856-57, p. 107. A male child, aged fourteen weeks, slight cyanosis. The ductus arteriosus is supposed to have been patent.

Baly, Path. Trans., vol. x., 1858-59, p. 90. Female infant, aged nine months, not decided cyanosis. For. ov. and duct. art. open. The pulmonary artery obliterated at its orifice.

Rokitansky, Wochenblatt der Zeitschrift, Jahrg. i., 1855, p. 225, quoted by Heine. See also the case of Fleischmann. In Tiedemann's Verengung, etc. p. 115, reference is made to a specimen in the Royal College of Surgeons of Edinburgh, in which the orifice of the pulmonary artery was obliterated from adhesion of the valves. The foramen ovale was largely open, and the bronchial arteries were also large. The condition of the left ventricle and duct. art. is not named. The preparation was obtained from a man aged twenty-two years, and does not now appear to be in the Museum.

Hervieux, L'Union Médicale, 1861, 9me année, 1861, p. 421. In this instance, the pulmonary artery is said to have been wanting, and the ductus arteriosus could not be detected. There were only three pulmonary veins ; the septum of the auricles was entire, but that of the ventricles defective. From a cyanotic child, aged three months.

[1] Dr. Paget on Congenital Malformations.—Edin. Med. and Surg. Jour., vol. xxxvi. 1831, p. 289.

duplicity of the right ventricle, consisting in the more or less decided separation of the sinus from the infundibular portion of that cavity, is by no means of uncommon occurrence. In this way we may explain the remark of Andral,[1] that he has seen a heart with three auricles, and another with four ventricles.

The well-formed human heart always displays a more or less marked separation between the two portions of the right ventricle, caused by the band of muscular fibres to which the cords of the tricuspid valve are attached; this condition indicating the mode in which the right ventricle is developed, as illustrated by the arrangement of the heart in the chelonian reptiles. Thus in the turtle the heart consists of three imperfectly separated ventricles; the right and left systemic ventricles, from which the two aortæ arise, and a small anterior ventricle which gives origin to the pulmonary artery. The latter is entirely separated from the left, but communicates with the right aortic ventricle. The sinus and infundibular portion of the right ventricle are, in man, the analogues of the right systemic and pulmonic ventricles of the turtle; and it has been pointed out by Mr. Grainger that the form of malformation here referred to, is to be ascribed to irregular development of the two portions of which the ventricle originally consists, by which the separation between them becomes much more decided than it should be.

Such separation may be produced in different ways. It may depend simply on undue development of the ordinary muscular bands, or on this in conjunction with thickening of the endocardium and subjacent fibrous tissue. The aperture of communication between the two portions of the ventricle may vary in size and form. In some cases, as when the partition is chiefly muscular, it may be of considerable size and irregular shape; in others, and especially when the endocardium and fibrous tissue also are thickened, the opening may have the form of an oval or round, firm, fibro-cartilaginous ring. In the latter case it not unfrequently

---

[1] Pathological Anatomy, translated by Townsend and West, vol. ii. p 333.

displays old or recent deposits of lymph upon its edges; and it may be so small as only to admit the passage of a crow- or a goose-quill, or there may be two or three small openings. In cases in which the obstruction caused by the septum is great, the sinus of the ventricle becomes large and its walls thick and firm, while the infundibular portion is compara- tively small and its walls thin and flaccid.

The defect may occur at different periods of fœtal life—at the early period when the septum is incomplete, and at the latter period when the heart is otherwise fully and naturally developed. In the former case the analogy to the form of the heart in the chelonia is most decided, the aorta arising from the one cavity,—the sinus of the ventricle, and the pulmonary artery from the other—the infundibular portion or conus arteriosus. In some cases the condition is combined with defective development at the outlet of the ventricle, situated either in the fibrous zone or in the valves or in both; and, as in such cases the infundibular portion of the ventricle becomes atrophied, the apparently separate cavity from which the pulmonary artery arises, may be reduced to a very small size.

One of the earliest instances in which this condition of the heart appears to have been noticed, was in a specimen preserved in the museum of St. Bartholomew's Hospital, and described by Dr. Farre.[1] There does not seem to have been any history of the case; but the subject of the malfor- mation was supposed to have been about fourteen years of age. The aorta arose from both ventricles; and the pul- monary artery was correctly formed and of natural size, and is stated to have arisen from a very small third ventricle, which communicated by two apertures with the right ven- tricle. A heart presenting a similar condition has been described and figured by Mr. G. C. Holmstead,[2] as found in a girl nine years of age; and Dr. Crampton[3] has reported a

---

[1] On Malformations, p. 26. Mr. Lawrence's case.

[2] London Medical Repository, vol. xvii. 1822, p. 455. The description is accom- panied by a woodcut, from which it is evident that the specimen affords a very characteristic example of obstruction at the commencement of the conus arteriosus. It is said to have been preserved in Sir C. Bell's Museum in Windmill-street.

[3] Transactions of the College of Physicians of Dublin, N. S., vol. i. 1830, p. 34, quoted in Todd's Cyclopædia of Anat. and Phys., vol. ii. p. 634.

third case which appears to have closely resembled the other
two, except that the pulmonic orifice was destitute of valves,
and was only closed by a puckering of the lining membrane
which occasioned some contraction. The defect occurred in
a boy ten years of age. Dr. Elliotson[1] has referred to a
case probably of a similar kind; and Dr. Theophilus Thompson[2] described another, in which the patient was a female
38 years of age, and the pulmonary orifice was wide and
provided with four well-formed and equal-sized valves. A
case has also been described by M. Raoul Chassinat,[3] in
which there appears to have been a septum in the right ventricle, while the pulmonary orifice was entirely obliterated;
the aorta arose from both ventricles and the branches to the
lungs received blood through the ductus arteriosus. The
malformation occurred, with other irregularities which have
before been referred to, in a child which lived twelve days.
Examples of this defect have also been published by M.
Aran and M. Deguise,[4] in a case which occurred in the
practice of M. Honoré; and by M. Pize;[5] and probably a case
published in Hufeland's Journal[6] may have been of a similar
description. In 1847 Mr. Le Gros Clark exhibited, at the
Medico-Chirurgical Society,[7] the heart of a young man,
nineteen years of age, which presented this malformation.
He had laboured under symptoms of cardiac disease, with
much lividity of the face, probably throughout his life. The

[1] Lumleyan Lectures, 1830, p. 21.

[2] Med.-Chir. Trans., vol. xxv. 1842, p. 247.

[3] Arch. Gén. de Méd., 2me série, t. xi. 1836, p. 81.

[4] Lancet, 1844, vol. i. p. 501; and Bullet. de la Soc. Anat., 17me année, 1842,
p. 180, and Thèse de la Cyanose Cardiaque, etc. Paris, 1843. Obs. 41. In a female
twenty years of age, who had suffered from the usual symptoms of malformation
from birth.

[5] Thèse de la Fac. de Paris, 1854, No. 148, obs. 29, p. 31. In a man twenty-
two years of age.

[6] Arch. Gén. de Méd., 2me série, t. ii. p. 101. In a female of twenty-five
years of age.

[7] Med.-Chir. Trans., vol. xxx. 1847, p. 113.

It is possible also that a case described by Nasse, Leichenöffnungen, 1821,
p. 116, may have been of this description. The septum of the ventricles was
entire but the foramen ovale open.

orifice of the pulmonary artery was provided with only two valves but it was not contracted, and immediately below the origin of the vessel there was a cavity of small size, partially separated from the other portion of the right ventricle; the only communication between the two cavities being through two small circular apertures, "neither of which would admit, without distension, the passage of a small-sized goose-quill." The circumference of the opening was dense and white, similar in appearance to the auriculo-ventricular zone. The septum ventriculorum was deficient at the base; so that a communication existed between the sinus of the right and the cavity of the left ventricle, by which the blood must have passed from the right ventricle into the aorta.

Shortly after Mr. Clark's case was read at the Medico-Chirurgical Society, I communicated a very similar instance of malformation which fell under my notice at the Royal Free Hospital. The following are the particulars of this case :—

CASE VII.[1]—*Constriction at the commencement of the infun-dibular portion of the right ventricle and at the orifice of the pulmonary artery ; aorta arising above an aperture in the septum ventriculorum ; death from embolism of the trunk and branches of the pulmonary artery.*

William Holland, a milk-boy, aged fifteen, was admitted under my care into the Royal Free Hospital, on the 20th of February, 1847. He stated that on the 13th he had bruised the left knee by a fall, and had since suffered from constant pain in that joint, and also, for two or three days before admission, in the right knee. The day after the accident he began to experience pain in the left side of the chest, and difficulty of breathing.

When admitted, he was much collapsed, and the extremities were cold and livid. The cheeks were of a deep

---

[1] Med.-Chir. Trans., vol. xxx. 1847, p. 131. The preparation of this case is contained in the Victoria Park Hospital Museum, and is numbered B 5. It is engraved in plate 6, fig. 2.

purple colour, and the lips blue. The fingers and toes were club-shaped, and the nails incurvated, and very dark coloured. The pulse was 124 in the minute and extremely feeble. The tongue was dry and covered with a whitish fur. The respiration was peculiarly rapid and panting, and he was compelled to lie on the back, partly inclined towards the right side, and with his head low. He complained of pain in the region of the heart, palpitation, and difficulty of breathing. Both knee joints were swollen and tender, and there existed a red and swollen patch over the left trochanter.

The chest yielded a clear sound on percussion, except in the præcordial region, where the resonance was impaired over a larger space than usual. The liver could be felt extending a little below the edges of the ribs. The sternum was arched and prominent, more especially towards its base. The respiration was puerile in character and attended with occasional mucous râles. A loud murmur was heard accompanying the impulse of the heart. It was of a soft or blowing character, and was most intense at the cartilage of the third left rib near the sternum, or at a point half an inch above the nipple and between that body and the sternum; the second sound was there inaudible. From this point it continued to be heard very distinctly though decreasing in intensity, along the upper part of the sternum, in the subclavian and carotid arteries, and on the left of the spine in the interscapular and dorsal regions. It was also heard less distinctly in a line from between the nipple and sternum, towards the middle of the left clavicle. Below the level of the nipple the murmur became shorter and more feeble, and, at the point of pulsation of the apex, towards the epigastrium, and on the right side of the lower half of the sternum, it was followed by a very clear second sound. The boy was much exhausted, and his intelligence so impaired that it was found impossible to collect any satisfactory information of his state of health previous to the present attack. He stated, however, that he had lost flesh and strength, and had been very subject to affection of chest, and of a livid complexion, since he was thrown from a cart twelve months before.

These notes were taken about 4 p.m. on the day of admission, and, notwithstanding the free use of stimulants, externally and internally, he gradually sank, and died at 8 o'clock on the following morning. On inquiry, I ascertained that he had always been of a somewhat livid complexion, but was stout, healthy, and capable of a full amount of exertion, till the occurrence of the accident he referred to. He was then riding at the back of a cart when it toppled up and he was thrown out and fell upon his head. He was admitted into the Royal Free Hospital on the 6th of February, 1846, under the care of my then colleague, Mr. Gay, with symptoms of concussion of the brain. He continued there six days, and the only peculiarity observed in his appearance was some slight lividity of the lips. Since that time he had been gradually getting thinner and weaker, and was constantly chilly and very subject to take cold. He complained occasionally of palpitation, difficulty of breathing, and pain in the region of the heart; and his hands and face were always very blue, but especially so in cold weather or when he was suffering from affection of the chest. His appetite was generally defective, and he occasionally vomited his food. His father is of a livid complexion, and has a " pigeon breast."

The *post-mortem examination* took place at 4 p.m. on the 22nd, thirty-two hours after death.

The brain was healthy, though much congested. It weighed 49 oz. 4 drachms avoirdupois.

The surfaces of the pleura on the right side were adherent by a small cellular band. The left lung was entirely free. Both lungs were engorged with blood, sparingly crepitant, and contained several masses in the state of pulmonary apoplexy. The bronchial mucous membrane was somewhat reddened.

The pericardium was healthy. The heart weighed 10 oz. It was broader from side to side than from above downwards. Its total circumference was 8½ French inches (229·5 mm.

9·05 ᴇ. in.), of which the right ventricle constituted 4½ (121·5 mm. 4·795 ᴇ. in.). The systemic veins were natural. The right auricle was large, and its walls thick. The foramen ovale was closed, with the exception of a small valvular opening capable of admitting a goose-quill. The Eustachian valve was of moderate size. The right auriculo-ventricular aperture measured thirty-nine lines in circumference (87·7 mm. 3·46 ᴇ. in.), and the valves were natural. The muscular column to which the cords of the anterior fold were attached, was very large and firm. The aperture opened as usual into the sinus of the right ventricle ; but this portion of the cavity was separated from the infundibular part by a thick muscular septum, defective only at its centre, over a space sufficient to admit the forefinger, and perforated by one or two small pores near the apex. The former cavity communicated with the aorta by an orifice thirty lines in circumference (67·5 mm. 2·66 ᴇ. in.), situated at its upper side. Its walls averaged 5½ French lines (12·37 mm. ·48 ᴇ. in.) in thickness, and were unusually firm and solid. In places they had undergone the fibro-cartilaginous degeneration throughout their whole extent, and the serous covering externally was opaque. The second or infundibular portion of the ventricle was of smaller capacity than the sinus, and gave origin as usual to the pulmonary artery. Its walls averaged only two or three lines in thickness. The orifice of the pulmonary artery was very small, and was provided with only two valves, which were extremely thick and opaque. The aperture on the ventricular side admitted a ball measuring fifteen lines in circumference (33·7 mm. 1·33 ᴇ. in.). The valves projected into the cavity of the vessel, leaving deep sacs behind them, and by their free borders occasioned further contraction of the orifice, so that it only gave passage to a ball of thirteen lines in circumference (29·2 mm. 1·15 ᴇ. in.).

The coats of the pulmonary artery were much indurated and thickened, and its canal was entirely obstructed by fibrinous coagula. At the sides of the vessel these coagula were of a dirty white colour, and were laminated and firmly adherent to the valves and lining membrane, but towards the centre of the canal they were softer and less decolorized. The obstruction occupied the whole trunk of the artery and extended a few lines into each of its branches. The smaller vessels were free from disease. The ductus arteriosus was impervious throughout the largest portion of its extent, but had a conical cavity extending two or three lines from the bifurcation of the pulmonary artery. The pulmonary veins were natural. The left auricle was small, and its lining membrane opaque. The left auriculo-ventricular valves were healthy, and the aperture measured thirty-six lines in circumference (81· mm. 3·19 E. in.). The left ventricle was of small capacity; its walls felt flaccid, and were 3½ to 4½ lines (7·87 to 10·12 mm. ·31 to ·39 E. in.) thick. The opening from the left ventricle into the aorta was of about the same size as that by which the vessel communicated with the right ventricle.

The aorta was healthy and very large from its origin to the insertion of the ductus arteriosus; from that point its calibre greatly diminished. The valves were of the usual number, and entirely free from disease. The bronchial and œsophageal branches were somewhat large. The veins and cardiac cavities were distended with blood.

The abdominal organs displayed no appearance of disease, but were much engorged. The liver weighed 45 oz., the spleen 6½ oz., the left kidney 4½ oz., and the right 3 oz. The latter was somewhat mottled.

In the following case the constriction at the commencement of the infundibular portion of the ventricle was very great, and the latter cavity was reduced to a very small space immediately below the pulmonic valves. The sinus of the ventricle was, on the contrary, very large, and the aorta arose in great part from it :—

CASE VIII.—*Great Contraction at the commencement of the infundibular portion of the right ventricle; aorta arising chiefly from the right ventricle; foramen ovale closed.*[1]

The specimen, which was sent to me by Mr. Roper, of Islington, was removed from a boy of seven years of age, who from shortly after birth had been cyanotic,— his cheeks, lips, and hands, being decidedly blue. He was also subject to syncopic attacks, attended by difficulty of breathing and palpitation. He died suddenly in one of these attacks, which occurred after he had eaten heartily of fruit tart.

The heart was of large size for the age of the subject, and when recent would probably have weighed about 6 oz. avoirdupois. The right ventricle was much increased in capacity, and its walls thicker than natural; the left ventricle being disproportionately small. The former measured externally in circumference 5·06 E. in. (128·2 mm.), the latter 3·19 E. in. (81 mm.). The aorta arose in great part from the right ventricle, the opening into the vessel from that cavity admitting of the passage of a ball measuring in circumference 2·66 E. in. (67·5 mm.), while the communication with the left of ventricle only allowed of the passage of a ball of 2·13 E. in. (54 mm.). The septum of the ventricles was largely deficient at the base. The commencement of the infundibular portion the right ventricle was very much contracted, being reduced to a round aperture, only allowing of the passage of a crowquill. This opening was situated a short distance below the attached margins of the semilunar valves, and left a small space between it and the valves which represented the infundibular portion of the ventricle or the conus arteriosus.

The constriction was partly due to hypertrophy of the muscular structure, and partly to thickening of the endocardium. The orifice of the vessel was small, but much larger than the opening below, and the valves were not

---

[1] Path. Trans., vol. xvii. 1865-66, p. 45.

materially thickened. They were only two in number, one of them being larger than the other, and that curtain displayed the appearance of consisting of two united segments. The valves were protruded forwards in the course of the artery so as to have deep sinuses behind them. The pulmonary artery was much less than the aorta. The aortic valves were considerably thickened ; the trunk of the vessel was of full size. The right auricle was somewhat dilated ; the foramen ovale completely closed. The mitral and tricuspid valves were not materially diseased. The aorta and pulmonary artery were cut off below the points at which the ductus arteriosus is given off and inserted, and the condition of that passage could not in consequence be ascertained.

In this case the foramen ovale was closed, and this was explained by the extremely defective state of the septum of the ventricles, by which a ready passage for the blood from the right to the left side of the heart was allowed. From the very small size of the opening from the right ventricle into the pulmonary artery, compared with the calibre of the vessel itself, it is most probable that the ductus arteriosus was not closed. Thus a further portion of the blood may have entered the pulmonary artery from the aorta, and have so been subjected to the influence of the air in the lungs. The parts having been cut short before I received the specimen, the state of the duct could not however be ascertained.[1]

Since the first of these cases was published I have had the opportunity of examining a specimen which exhibits the form of malformation in a very aggravated degree, and in conjunction with great defect in the conformation of the heart. The preparation was shown to me by Mr. Hutchinson.[2] It was removed by Mr. Keyworth of York, from a cyanotic girl about twelve years of age, who had been extremely susceptible to cold and incapable of any active exertion throughout her life. The septum between the sinus and infundibular

---

[1] The preparation is contained in the Museum of the Victoria Park Hospital, and is numbered B 26.

[2] Path. Trans., vol. v. 1854, p. 99.

portion of the right ventricle is constituted as in the first case, by interlacing muscular columns, and the aperture is only sufficiently large to allow the passage of a small probe, the contraction being increased by fibrinous deposits on the edges. The infundibular portion of the right ventricle forms a cavity about eight lines in length, between the septum and the orifice of the pulmonary artery. The pulmonary artery is somewhat small, and its valves are apparently only two in number. The sinus of the right ventricle is a small cavity communicating above with the right auricle, and separated from the left ventricle only by muscular bands; indeed, it seems to constitute a portion of the latter cavity. The left auricle opens naturally into the left ventricle. That cavity is very large, and its walls thick. It gives origin to the aorta, and has also an indirect communication with the right auricle. The aorta is of large size. The foramen ovale is entirely closed, and the ductus arteriosus is pervious some distance from the pulmonary artery, but becomes obliterated before its union with the aorta. The auriculo-ventricular and aortic valves are much thickened.

In all the cases just referred to, the development of the supernumerary septum occurred before the division of the ventricles was completed, or at an early period of fœtal life; but this rule does not apply to other instances in which similar septa have been found. In the following case the abnormal partition constituted the only defect in the development of the heart.

CASE IX.[1]—*Constriction at the commencement of the infundibular portion of the right ventricle; heart otherwise well formed.*

A female child, five years of age, was brought under my notice on the 20th of December, 1846. She was stated to have been livid from birth, but acquired a more natural

---

[1] Med.-Chir. Trans., vol. xxxi. 1848, p. 61. The preparation of this case is contained in the Victoria Park Hospital Museum, and is numbered B 2. It is engraved in plate 6, fig. 1.

colour soon after, and was stout and healthy till between two and three years of age; she then began to suffer from difficulty of breathing without any assignable cause, had a slight cough, and became thinner. Since that period she had continued delicate, and was very susceptible to cold. When chilled or suffering from catarrhal symptoms, she became livid in the face, and had hurried and difficult breathing. She was repeatedly seen and examined during the year 1847. At that time her cheeks were tumid and much flushed, the vessels being distinctly visible. The arms and hands were puffy, and the fingers and toes club-shaped at their extremities, and of a deep red or purple colour, but not blue. The pulse was always more or less accelerated. There was some difficulty of breathing, and a slight hacking cough. The dull space at the præcordia was somewhat greater than natural, but the chest was elsewhere fully resonant. Slight sibilant and sonorous râles were heard with the respiration; and over a large portion of the front of the chest a loud systolic murmur was audible. This murmur was thought to be most intense midway between the left nipple and sternum; but it was also very distinct from this point towards the middle of the left clavicle, across the sternum to the right side, and along the whole of the middle and lower part of the sternum. In these situations the diastolic murmur was indistinct, but, at the upper part of the sternum and at the point of pulsation of the apex of the heart, the murmur was less intense and prolonged, and the diastolic sound clear. A feeble murmur was audible to the left of the spine, in the interscapular region. There was no permanent turgidity or pulsation of the jugulars. The liver was large, and the abdomen tumid. In September, 1847, the child returned from the sea-side, where she had been for six weeks or two months, greatly improved in general health. The murmur, however, though less intense, was still audible, and the lividity equally marked. Soon after this she took scarlet fever, had severe ulceration of the throat, followed by vomiting of blood in large quantities, and died exhausted in about three weeks.

The body was examined on the 2nd of November. It was much emaciated. There were slight old adhesions at the lower and posterior part of the left lung, and some lobular condensation in both lungs. The smaller bronchial tubes contained a little secretion, but the mucous membrane was not materially reddened. The heart was of natural form. It weighed 3¾ oz. avoirdupois. There was a slight deposit of fat on the surface of the right ventricle, and some old adhesions between the aorta and pulmonary artery. The right auricle was large, and distended with imperfectly coagulated blood. The foramen ovale was completely closed. The right auriculo-ventricular aperture admitted a ball measuring in circumference thirty-nine French lines (87·75 mm. 3·46 E. in.) : the valves were natural. In the cavity of the right ventricle there existed a septum, dividing the sinus from the infundibular portion; and this septum was perforated by an oval aperture, twenty-one lines (47·25 mm. 1·86 E. in.) in circumference, by which the two divisions of the cavity communicated. The edges of the aperture were smooth, and on the auricular side the lining membrane and muscular structure around had undergone the fibro-cartilaginous transformation. The walls of the sinus of the ventricle had an average thickness of two lines ; those of the infundibular portion, of only one line. The pulmonary orifice had a circumference of 26½ lines (59·62 mm. 2·35 E. in.). Its valves were natural, and the ductus arteriosus was occluded. The left cavities were natural ; but the left auriculo-ventricular aperture and the orifice of the aorta, were smaller than the corresponding orifices on the right side. The liver and spleen were large. Both kidneys were extensively diseased, mottled with purple patches, and very lacerable.

A very similar case to that which has just been related has recently been published by M. Claude Bernard,[1] which occurred in a female 56 years of age, who had suffered from an attack of acute rheumatism twenty years before,

---

[1] Arch. Gén. de Méd., 5ᵐᵉ série, t. viii. 1856, p. 167.

and died with the usual symptoms of cardiac asthma and dropsy. On examination, the left side of the heart was found healthy, but there was great hypertrophy of the right ventricle; and, in the infundibular portion of the cavity there was a ring composed of firm, resistent fibrous tissue, which would admit the point of the little finger, and had a diameter of only 10 or 12 millimetres ($\cdot$39 to $\cdot$47 E. in.). The septum was situated at least a centimetre ($\cdot$39 E. in.) below the valves of the pulmonary artery. The orifice and valves of the artery were natural but its trunk was dilated, and the heart presented no other defect except the general enlargement. M. Bernard ascribes the production of the constriction, in this case, to endo-carditis occurring at the time of the rheumatic attack. From, however, the similarity of the defect to that which existed in the former instances, and from the circumstance that in the girl whose case has just been related the symptoms had been noticed from early life, and that in the other instances there were undoubted malformations, it is more probable that the ring was due to irregularity of development. The obstruction may have been materially aggravated after the rheumatic attack, as the result of increased contraction from thickening of the endocardium. In reference to one of the other recorded cases, it has been suggested by the narrator, that the septum was due to hypertrophy of the muscular columns consequent on constriction of the pulmonic orifice. There was, however, no obstruction at the pulmonic aperture in either of the latter cases, or in those of Mr. Holmstead, Dr. Theophilus Thompson, and Mr. Le Gros Clark; in all of which the orifice and valves of the pulmonary artery were healthy. The only explanation applicable therefore to the whole of the cases of this description, is that they are examples of irregular development, the deviation from the natural conformation taking place at different periods of foetal life.

The preparation of Mr. Clark's case is contained in the museum of St. Thomas's Hospital, and is numbered LL 63. There is also in the museum another preparation, which affords a very good example of this description of malformation combined with defect in the septum. It is numbered LL 72, and is stated to have been removed from a child.

Since the last edition of this work appeared, a very interesting instance of this description of anomaly has been published by Dr. Kussmaul, of Freiberg, in an elaborate and able memoir, in which he has thrown much light on the nature of the defect and the circumstances under which it occurs, to which I have been indebted in the revision of this article. The case occurred in a girl twelve years of age, who had been subject to attacks of profuse bleeding from the nose, but was not cyanotic, and died after an illness of five days, characterized by pain in the præcordial region, &c. The conus arteriosus was connected with the sinus by a small slit, and the communication with the pulmonary artery was also greatly contracted. The ventricular septum was imperfect, and the foramen ovale was open, but the ductus arteriosus was obliterated. In the case of Dr. Dickenson, also, to be hereafter referred to, there was, in addition to transposition of the cavities of the heart and the origin of both of the large arteries from the right ventricle, a very marked constriction at the commencement of the conus arteriosus. That part of the ventricle was indeed reduced to a very small cavity from which the pulmonary artery arose.[1]

4. *Defect in the inter-ventricular septum from constriction of the auriculo-ventricular and aortic apertures.*

In most of the cases which have been mentioned, the defective condition of the septum of the ventricles was associated with, and probably dependent on, obstruction at the pulmonic orifice, or at the point of union of the infundibular portion and sinus of the right ventricle. It is probable that the septum may equally remain unclosed in consequence of disease of the right auriculo-ventricular aperture and of the aortic or mitral orifices. In the *Lancet* for 1848, a case is related by Mr. Robinson,[2] in which the right auriculo-ven-

---

[1] See also Förster, Taf. xix. figs. 11, 12, 13, and 14.　　[2] Vol. ii. p. 103.

tricular valves were diseased and the septum of the ventricles imperfect, in a child a year and a half old, without any defect at the pulmonic orifice. In a case recorded by M. Burguières,[1] the septum cordis was also found imperfect in a girl of nineteen, in conjunction with some disease of the mitral valve and extensive obstruction at the aortic orifice, caused by the union of the valves into a diaphragm, leaving an aperture only fourteen lines (31·5 mm. 1·24 E. in.) in circumference.

In the following case, the defect in the inter-ventricular septum co-existed with disease of the tricuspid valve.

CASE X.[2]—*Two small apertures in the septum ventriculorum ; contraction of the right auriculo-ventricular aperture ; foramen ovale and ductus arteriosus closed.*

A female infant was brought to me at the Hospital for Diseases of the Chest, on the 1st of March, 1853.　It was then three months old, and was stated to suffer from paroxysms of difficulty of breathing, during which the face and upper extremities became very dark, and the heart beat violently.　The child was unusually livid when born ; but it subsequently acquired a more natural colour, and then at the end of the first week again became darker.　The palpitation and dyspnœa were observed during the second week, and those symptoms continued till she was brought to the hospital.

On the 2nd of March the following notes were taken :— " The child has had no severe paroxysms since she was last seen ; but the hands and face have been constantly more or less livid ; there is, however, now no blueness beneath the nails.　The mother states that the left arm and hand are always of a darker hue than the right, and that the limb is somewhat swollen, and such is certainly the case at present.　The lower extremities are of the natural colour. The action of the heart is powerful, and there is a systolic

---

[1] Thèse de la Faculté, Paris, 1841 ; quoted by Deguise, Thèse, 1843.

[2] Path. Trans., vol. v. 1853–54, p. 64.　The preparation is contained in the Museum of the Victoria Park Hospital, and is numbered B 10.

murmur audible over the whole front of the chest, but most intensely in the præcordia and to the left of the lower part of the sternum. It is also very distinct at the upper part of the sternum and beneath the left clavicle, but is less intense on the right side of the sternum. Behind, it is only indistinctly audible; but it is somewhat louder to the right than to the left of the spine. There is some dulness on percussion in the dorsal regions, especially the left. Bronchitic rhonchi are heard in all parts of the chest. The pulse is regular and equal at the wrists. The child is very restless and becomes extremely livid when excited, as by crying, or by the slightest exposure to cold." From this date the paroxysms were less frequent and severe, and the child had generally a more natural appearance; but it did not thrive, or rather it became more emaciated.

In the beginning of June the child took hooping-cough, under which other members of the family laboured at the time. The symptoms then became much aggravated; the breathing was rapid and laborious, and the action of the heart tumultuous. During the fits of coughing the face and extremities were intensely livid—almost black—and suffocation seemed impending. It died convulsed on the 29th of June, when about seven months old.

The body was much emaciated. There were adhesions between the two layers of pericardium at the apex of the heart. The heart weighed 2½ oz. avoirdupois. Its form was somewhat more pointed than usual; the apex was formed by the left ventricle. Both auricles contained coagula, and the left was very much distended by firmly coagulated but dark-coloured blood. The right auricle was of ordinary size; the Eustachian valve distinct. The right ventricle was small, but its walls were thick and dense. The pulmonary artery was of unusually large size, its orifice admitting of the passage of a ball twenty-one French lines (47·25 mm. 1·86 E. in.) in circumference. The folds of the right auriculo-ventricular valve were thickened and adherent at their angles, so as to contract the dimensions of the orifice, which only admitted a ball measuring twenty-four lines (54 mm. 2·13 E. in.) in circumference. On the auricular surface of the valve there was a thick exudation of recent lymph. The cavity of the left auricle was large, and the walls thicker than usual. The foramen ovale was closed, but the fossa was depressed towards the left side. The left auriculo-ventricular aperture had a circumference of eighteen lines (40·5 mm. 1·59 E. in.). The walls of the left ventricle were less firm than those of the right. The aorta arose from the left ventricle, but there were two openings in the septum ventriculorum by which the ventricles communicated; these apertures were much larger on the left than on the right side. Both openings led into the right ventricle, behind the auriculo-ventricular valves, and the largest had a circum-ference of six lines (13·5 mm. ·53 E. in.). The ductus arteriosus was completely closed.

The glands at the root of the lungs and in the course of the large vessels were much enlarged. The thymus gland was large. The lungs were in some places collapsed, in others emphysematous; much viscid mucus was contained in the smaller bronchial tubes in the collapsed portions of the lungs. The liver, spleen, and kidneys were engorged, but otherwise healthy.

5. *Obliteration or Atresia of the auriculo-ventricular and aortic apertures.*

Closely allied to the forms of malformation which have just been alluded to, are those in which other passages of the heart, which should be permanent, are found wanting. Of obliteration of the pulmonary orifice I have already spoken. The absence of the right auriculo-ventricular aperture has also been incidentally mentioned as noticed in one of the cases described by M. Thore,[1] in that of M. Claude Bernard, and in a preparation in the Museum of St. Thomas's Hospital.[2] In the case of M. Valleix,[3] and one of partial transposition of the arteries, veins,

---

[1] P. supra, 25.   [2] P. 25.   [3] P. 96 infra.

and auricles, related by Dr. Worthington,[1] which occurred in a female child twenty-two months old; and in a case described by Dr. Favell,[2] in which the pulmonary artery was found without valves and the foramen ovale was largely open, in a boy eight years old, the right auricle and ventricle also had no connexion. A similar condition existed in the heart of a child which lived nine weeks, exhibited at the Pathological Society, by Dr. Sieveking, in 1853.[3] The septum of the ventricles was defective, the foramen ovale was largely open, and the right ventricle which gave origin to the pulmonary artery, had no communication with the right auricle. The same condition was met with by Vrolik, in a cyanotic child nine years of age,[4] and occurred in the case of Mr. Holmes to be hereafter mentioned.

Of obliteration or absence of the left auriculo-ventricular aperture, the cases of Professor Owen and Mr. Clark,[5] and Dr. Vernon,[6] afford examples. A similar defect existed in a heart exhibited at the Société Anatomique of Paris, by M. Parise,[7] and in a case which occurred at Strasburg and is reported by M. Valette.[8] In a case related by Dr. Blackmore,[9] the left auriculo-ventricular aperture is said to have been obliterated, but a fissure existed by which the left auricle and ventricle communicated. A preparation exhibiting the defect is also stated to exist in the Museum at Göttingen, which occurred to Professor Hasse.[10]

Obliteration of the aortic orifice is reported to have been found by Romberg,[11] in a child which lived four days, and was completely cyanosed. The right ventricle was dilated

---

[1] American Jour. of Med. Sc., vol. xxii. 1838, p. 131.

[2] Provincial Medical Journal, vol. iii. 1842, p. 440.

[3] Path. Trans., vol. v. 1853–54, p. 97. This specimen is preserved in the Museum of St. Mary's Hospital.

[4] Förster, Taf. xviii. fig. 17–18.   [5] Supra, p. 16.

[6] Supra, p. 17.   [7] Bullet. de la Soc. Anat., 12me année, 1837, p. 100.

[8] Gaz. Méd. de Paris, 1845, p. 97.

[9] Edin. Med. and Surg. Jour., vol. xxxiii. 1830, p. 268.

[10] Ibid., Taf. xviii. fig. 10–12.

[11] Tiedemann's Verengung und Schliessung der Pulsädern. Heidelberg and Leipzig, 1843; of which an analysis is given in the Edin. Med. and Surg. Jour., vol. lxv. 1846, p. 149.

and hypertrophied, and the pulmonary artery large. The left auricle and ventricle were very small, and there was not a trace of the aortic orifice. The foramen ovale was largely open, and the supply of blood to the aorta was conveyed from the pulmonary artery by the ductus arteriosus. An example of the same defect was exhibited at the Pathological Society by Mr. Canton,[1] in 1849; and by the kindness of that gentleman, I have had the opportunity of examining the specimen. The canal of the aorta extends nearly up to its origin, but the vessel there becomes entirely impervious, apparently from the adhesion of the valves, so that there is no communication between the artery and the cavity of the left ventricle. The pervious portion of the aorta received its supply of blood from the right ventricle through the ductus arteriosus and pulmonary artery. The septum of the ventricles is entire, and while the right ventricle is very large, the cavity of the left ventricle is nearly obliterated. The foramen ovale is freely open and the left auricle is natural. The subject of the defect was a child which lived two days, and was apparently healthy till seized with the convulsions in which it died. According to Förster a specimen exhibiting the same defect exists in the Pathological Museum at Würzburg. It was removed from a child nine days old.[2]

6. *Defect in the Inter-ventricular Septum; Pulmonary Artery arising from the left Ventricle.*

It has been already shown that the form of defect in which the inter-ventricular septum is imperfect at the base and the aorta arises wholly or in part from the right ventricle, is of frequent occurrence. The opposite condition, in which the deviation of the septum is to the right, so that the aorta and pulmonary artery both arise from the left ventricle, is, on the contrary, a very rare anomaly. A case which was related to the Académie des

---

[1] Path. Trans., vol. ii. 1848-49, 1849-50, p. 38.
[2] Förster, Taf. xix. fig. 2 and 3.

Sciences by M. Méry, in 1700,[1] was, however, probably of this description, though it has been quoted as an example of biloculate heart. The fœtus was of very defective conformation, the large cavities being open. The heart consisted of a common auricle, into which both the pulmonary veins and the venæ cavæ entered. This cavity communicated by a considerable aperture with the right ventricle, and by a smaller one with the left ventricle. The right ventricle had no artery arising from it, but opened into the left, which gave origin both to the aorta and pulmonary artery.

A case occurred to M. Maréchale,[2] which also affords an example of this kind of defect. Both the pulmonary artery and aorta arose from the left ventricle, and that cavity was of large size; while the right ventricle was rudimentary, and communicated with the right auricle and the left ventricle. The auricular cavities were imperfectly separated. The subject of the case was an infant, which attained the age of nearly four months, and presented the usual symptoms of malformation of the heart.

A case related by Mr. Holmes,[3] of Montreal, also presents a somewhat similar malformation. It occurred in a young man, who died at the age of twenty-one, after having, throughout his life, laboured under palpitation, dyspnœa, and blueness of the lips, terminating in dropsical symptoms. The two auricles both opened into the left ventricle, and the right ventricle had no connexion with the right auricle but gave origin to the pulmonary artery, and communicated with the left ventricle by an aperture in the septum. Since this case was published, another, resembling it in several respects, has been recorded by M. Valleix.[4] The subject of the malformation was an infant, which had double hare-lip, and died when about eight days old, after having undergone the operation. During its short life it presented no evidences of defect of the circulatory organs. The viscera of

[1] Hist. de l'Acad. des Sc., année 1700, Paris, 1703, obs. 17, p. 42.
[2] Quoted in Gintrac, sur la Cyanose (obs. 46, p. 173), from the Jour. Général de Méd., 1819.   [3] Edin. Med.-Chir. Trans., vol. i. 1824, p. 252.
[4] Bullet. de la Soc. Anat., année 9, 1834, p. 253.

the body generally were transposed. There were two auricular appendages, but only a common cavity. Into this cavity two descending cavæ opened, one on the right the other on the left side. The inferior cava entered on the left side, and the pulmonary veins on the right. The auricle opened into a large left ventricle by an aperture guarded by a tricuspid valve. The left ventricle gave origin to the aorta, and communicated with a second cavity which was as it were hollowed out of the ventricular walls and represented the right ventricle. This cavity gave origin to the pulmonary artery, and had no communication with the auricle except through the other ventricle. The ductus arteriosus was natural.

Since the last edition of this work was published, cases somewhat similar to those here referred to, have been published by Dr. George Buchanan, Dr. Kussmaul, and Dr. Dickenson.

In the case of Dr. Buchanan,[1] the child, a boy four years old, died of jaundice, having been deeply cyanotic during life. The pulmonary artery arose from the left ventricle, and was only about half its proper size, and the aperture was provided with only two valves. The aorta was connected with both ventricles, the septum being defective. The foramen ovale was open, but the ductus arteriosus was obliterated. In the case related by Dr. Kussmaul,[2] the child died when two years and three months old, having suffered from symptoms of cardiac disease but without cyanosis. The trunk of the pulmonary artery was found constricted, the conus arteriosus rudimentary, and the pulmonary artery and aorta both arose from the left ventricle. The foramen ovale, and the ductus arteriosus at its aortic end, were obliterated. The case of Dr. Dickenson, which was somewhat different from both the others, was reported to the Pathological Society during the present session. The subject of the defect was a male child who died at the age of three and a half years, and had been cyanotic from birth. The heart was situated on the

---

[1] Path. Trans., vol. viii. 1856–57, p. 149.
[2] Zeitschrift für Rationelle Medicin.—Henle und Pfeuffer, xxvi. band. 1865, p. 99.

right side instead of the left, and the aorta made its turn from left to right, giving off the innominate trunk on the left side and the separate carotid and subclavian arteries on the right. The auricles and ventricles were transposed : the pulmonary auricle opened into a small cavity, the analogue of the left ventricle, and this again communicated with a large cavity corresponding to the right ventricle, though placed on the left side, from which both the aorta and pulmonary artery arose. There was also a very marked constriction between the sinus and infundibular portion of the right ventricle. The sinus was large, and gave origin to the aorta, while the infundibular portion was reduced to a very small cavity, situated immediately anterior to the origin of the pulmonary artery.[1]

Cases of this kind afford examples of complex deviations from the natural process of development, consisting partly in arrest of growth at early periods of foetal life, and partly in irregularity in the evolution of the main arteries— the aorta and pulmonary artery,—from the single arterial trunk and the branchial arches.

III. MALFORMATIONS OCCURRING DURING THE LATER PERIODS OF FŒTAL LIFE.

1. DEFECTS PREVENTING THE HEART UNDERGOING THE CHANGES WHICH SHOULD ENSUE AFTER BIRTH.

PREMATURE CLOSURE OF THE FŒTAL PASSAGES.

IN some cases the heart has been found incapable of maintaining extra-uterine life, in consequence of the faulty development caused by the closure of the foramen ovale or ductus arteriosus in the foetus.

Premature closure of the Foramen Ovale.

The first recorded instance of this defect is that related by Vieussens, in 1715.[2]   The child which was the subject of

---

[1] Path. Trans., vol. xvii. 1865–66, p. 83.

[2] Traité de la Structure et des Causes du Mouvement du Cœur, ch. viii. p. 35 ; quoted by Corvisart, "Essai," troisième édition, 1818, p. 315.

the malformation lived thirty hours, and when born appeared well formed and healthy, but during its short life had difficulty of breathing, a leaden hue of the surface, and cold extremities. The right ventricle and pulmonary artery were extraordinarily developed, and there was no trace of the foramen ovale. Recently two other cases of this description have been recorded in this country. The first of these occurred in the practice of my friend Mr. E. Pye-Smith,[1] in a male child which was healthy looking at the time of birth, but within five minutes became livid and had difficulty of breathing. It then lapsed into coma and died in twenty-one hours. The fossa ovalis was in its natural situation, but was entirely closed by a reticulated membrane, which formed a pouch projecting towards the left auricle. There was a mere vestige of the Eustachian valve. The right auriculo-ventricular aperture, the right ventricle, and the pulmonary artery were of unusually large size; while the left auricle, and the left auriculo-ventricular and aortic apertures were very small, and the left ventricle was nearly obliterated. The ductus arteriosus was large and opened into the aorta somewhat more directly than usual.

The other case is reported by Dr. Vernon,[2] and is very similar to that of Mr. Pye-Smith so far as the obliteration of the foramen ovale was concerned; but the septum ventriculorum was deficient over a considerable space, so that the aorta arose from both ventricles. The right cavities, as in the former cases, were largely developed; the left, on the contrary, were small. The infant survived four hours and a half. It was very livid during life, and died in convulsions.

Premature obliteration of the Ductus Arteriosus.

The ductus arteriosus may also become obliterated at different periods of foetal life. At least it is by no means unusual in malformed hearts for no remains of the duct to

---

[1] Path. Trans., vol. i. 1846–57, 1847–48, p. 52.
[2] Med.-Chir. Trans., vol. xxxix. 1856, p. 299.

H 2

be found, and this may be referred to that portion of the branchial arch having become abortive at an early period. Thus the duct may be absent when the septum of the ventricles is very imperfect, though two distinct vessels arise from the cavity ; and it may also not exist in cases where the development of the organ is more advanced, the septum of the ventricles being only slightly defective. Of the former condition I have already quoted examples recorded by MM. Chemineau[1] and Thore ;[2] of the latter, cases related by Knox, Blackmore, Lexis, Huss,[3] Deguise, and Aran,[4] and Worthington,[5] afford instances.

When the ductus arteriosus is prematurely closed, the pulmonary artery gives passage during foetal life only to the very small amount of blood which circulates through the lungs. It is in consequence so imperfectly developed as to be incapable of adequately expanding after birth, and of conveying the larger quantity of blood which should then be transmitted. The small size of the vessel thus becomes a source of permanent obstruction, which entails other defects in the development of the organ. Such seems to have been the origin of the malformation in some of the instances referred to, and especially in the case of Dr. Ramsbotham related at page 66. The following cases also afford examples of the same kind :—

CASE XI.[6]—*Smallness of the pulmonary orifice and artery; absence of the ductus arteriosus; defect in the septum of the ventricles; foramen ovale open.*

The preparation of this case was obligingly forwarded to me by Dr. Bentley, and I am further indebted to him for the following particulars of the patient during life :—

When the child was first born, it was noticed to be very

---

[1] Page 21 supra.   [2] Page 18.
[3] See reference to these cases at pages 57–60.
[4] Page 78.   [5] Page 94.
[6] Path. Trans., vol. vii. 1856, p. 83. The preparation is contained in the Victoria Park Hospital Museum, and is numbered B 9.

livid, a fluttering was perceived at the heart, and the respiration was very imperfect. These symptoms continued for several days, after which it became less livid, but it never acquired the natural colour. About seven weeks after birth, the child began to suffer from attacks of extreme dyspnœa, during which its breathing was rapid and laborious; it threw its head back and gasped for breath; its complexion became almost black, and it cried incessantly. These fits subsided after loud stridulous breathing, and the child became cheerful; but they were brought on by the slightest excitement, and especially on taking the breast or food. At first the fits only occurred about once daily; but towards the close of life there were sometimes many in the day—on one occasion fully thirty. The sounds of the heart were not well observed but there was apparently no murmur; the beats were, however, rapid and violent, and the radial pulse weak. Till the last month the child was fairly nourished but it afterwards became extremely emaciated. It died when twelve months and three or four days old.

The child was born at the full period, and of healthy parents. It was their fifth child, and another was said to have something the matter with its heart.

On examination, the membranes and substance of the brain were found congested, and at the base there was recent lymph. There was a large effusion of serum in the ventricles and beneath the arachnoid. The lungs were much congested, and in some places solid as if unexpanded.

The heart weighed $2\frac{1}{2}$ oz. avoirdupois. The right ventricle was larger and its walls thicker and firmer than those of the left. The aorta arose from both ventricles, there being a large deficiency in the inter-ventricular septum. The pulmonic orifice was only eight lines (18 mm. ·73 E. in.) in circumference. The valves were two in number, and they were thick and indurated. They were protruded forwards into the artery, so as to form a funnel-shaped aperture from the ventricle, and to leave deep sinuses behind the folds. The pulmonary artery was throughout of very small size, and no traces of the ductus arteriosus could be detected. The foramen ovale was largely open, and the Eustachian valve was very perfect. The ascending aorta was of large size, and suddenly became smaller after giving off the left subclavian artery.

In reference to this case, it might be supposed that the primary deviation from the natural process of development was the imperfect condition of the pulmonic valves; but this is not probable. Obstruction of the pulmonic orifice, though it may occasion defects in the ventricular and auricular septa, so far from causing the premature obliteration of the duct, would rather occasion the persistence of that passage. It seems, therefore, most probable that the primary error consisted in the obliteration of the duct at an early period of fœtal life. The smallness of the pulmonary artery thus occasioned might entail the defect in the valves, or they may have been involved in the original error.

The following case affords an instance of the same kind of defect combined with irregularities in the course of the aorta and in the origin of the primary vessels:—

CASE XII.—*Absence of ductus arteriosus; small size of the pulmonary artery; aorta arising from both ventricles; irregular course of the aorta, &c.*

The subject of this case was a male infant, under the care of Dr. Brinton, at the Royal Free Hospital, to whom I am indebted for the opportunity of watching the case. He was born at the full period and of healthy parents. His mother stated, that he was when born a well-formed child, but three weeks after he was seen by a medical man, and found to be of an unusually dark colour. From this time the lividity increased, and at the age of seven or eight months, he began to suffer from suffocative attacks which terminated in convulsions. Of these he sometimes had several slight fits during the day; at other times the attacks were less frequent but more severe and of longer duration. When in them he became literally almost black in the face, and the hands and feet, and especially the nails, were quite black. When quiet the heart-sounds were distinct and free from murmur; but when he was excited, a loud systolic murmur was audible over the whole front of the chest, and most intensely at the middle and upper part of the sternum. The respiration was generally rapid and the pulse also quick. The veins of the head and neck were greatly distended. The child was excessively irritable, and seldom ceased crying or slept except when carried about in the arms. Towards the latter period of his life, he had frequent sickness and vomiting and constant diarrhœa. He became excessively emaciated and died exhausted after a severe fit, when eleven months and a half old.

On examination, the heart was found much larger than that of a healthy child at the same age. The pulmonary artery, which arose as usual from the infundibular part of the right ventricle, was of very small size and very short. It divided into two pulmonary branches, but there was no trace of the ductus arteriosus. The aorta arose partly from the sinus of the right ventricle, and communicated with the left ventricle by an aperture at the base of the septum. The right ventricle was large and formed almost the whole of the anterior part of the organ, and its walls were thick and firm. The left ventricle was small and its walls less firm. The right auricle was very large, the left auricle small, and the foramen ovale entirely closed. The aorta pursued an unusual course; it passed over the right bronchus, made its turn behind the termination of the trachea to reach the common situation on the left side of the bodies of the vertebræ, and thence followed its ordinary course. The branches arising from the arch were four in number. Shortly after the origin of the vessel, the right and left carotid arteries arose and passed upwards on each side of the front of the trachea. The right subclavian artery was then given off when the aorta was at the side of the trachea, and the left subclavian after it had turned round the right bronchus. The aorta then passed outwards, lying in a sulcus at the point where the left bronchus separates from the trachea, and followed its ordinary course. The bronchial arteries given off from the upper part of the descending aorta were much larger than usual.

The above case of malformation possessed several features of peculiar interest.

1. It probably originated in the imperfect development of the branchial arches, which usually go to the formation of the left aorta and ductus arteriosus. In consequence of the absence of the ductus arteriosus, the pulmonary artery received only the blood transmitted to the lungs during fœtal life, and continued permanently small; the septum of the ventricles was imperfect; and the aorta arose from the right ventricle. Indeed, the right aorta was developed instead of the left.

2. The aorta not only originated irregularly, but it pursued an unusual course passing over the right bronchus ; and it gave off four, instead of only three, branches from the arch.

In neither of these respects is the case unique. Instances of absence of the ductus arteriosus have already been referred to, and the defect has also been found in a monstrous child, by Otto;[1] and in a fœtus, by Meckel.[2] In some of these cases the duct probably became abortive from the free communication between the ventricles and the common origin of the aorta; in others, the closure of the duct might, as in this instance, be the primary change. It will be observed that the pulmonary artery was very short; and this is probably commonly the case when the ductus arteriosus does not exist.

The irregular course of the aorta, in which that vessel makes its turn round the right bronchus, has been met with in the case of malformation of the heart described both

---

[1] Selt. Beob. pt. i. p. 16.   [2] Reil's Arch., vol. ix. p. 437.

by Cailliot[1] and Obet,[2] and in a case where the heart was healthy, related by Sandifort[3] and Agliette,[4] and in one by Otto.[5] In the case described by Sandifort and Agliette, the vessels arising from the arch presented a very similar irregularity to that which existed in this instance; but, in addition, the obliterated trunk of the ductus arteriosus was found inserted at the commencement of the left subclavian artery. In the case of Cailliot and Obet, the vessels arising from the arch were transposed, the right subclavian and carotid arteries arising separately, and the left subclavian and carotid by a common trunk. In the case of Otto, five vessels were given off from the arch—the left carotid, the right carotid, the vertebral, the right subclavian, and the left subclavian. Walther has described and figured a specimen in which, with the natural conformation of the heart and the aorta in its normal position, four vessels were given off—viz., two carotid and two subclavian arteries;[6] and a similar irregularity was met with by Dr. Walshe,[7] in a case of transposition of the aorta and pulmonary artery. It appears by no means uncommon that the four primary vessels arise separately; but the arrangement is generally different from that which obtained in this case—the carotids arising first, then the left subclavian, and, lastly, the right subclavian, and the latter vessel usually passes behind the trachea to reach the right side of the neck. This irregularity has been described and figured by Boehmer;[8] and three instances of the kind exist in the Museum of St. Thomas's Hospital.

3. The cyanosis in this case was very intense, a pecu-

---

[1] Bullet. de l'École de Méd., 1807, No. 2, p. 24, p. 57 supra.

[2] Bullet. des Sc. Méd., t. ii. 1809, Mai, 1808, p. 65.

[3] Mus. Anat. Lugd. Batav., 1793; vol. i. 273; and vol. ii. plate 107, figs. 1 and 2.

[4] Saggi di Padova, 1786, t. i. p. 69, tav. 1.

[5] Neue Selt. Beob., 1824, p. 60; and Verzeichniss, 1826, No. 1922.

[6] Nouv. Mém. de l'Acad. de Berlin, 1785, plate 61, xiv.; and Tiedemann, tab. 3, fig. 5; and Quain and Maclise, plate 6, fig. 12.

[7] Med.-Chir. Trans., vol. xxv. 1842, p. 1.

[8] Haller's Disp. Anat., vol. ii. 1747.

liarity which may be explained by the difficulty which must
have existed in the transmission of the blood through the
lungs and systemic vessels.   The double circulation was
almost wholly maintained by the right ventricle; and the
small portion of blood conveyed to the lungs by the contracted pulmonary artery, must have been returned by the
pulmonary vein to the left auricle, thence transmitted into
the left ventricle, and so, by the aperture in the septum,
into the aorta.   The closure of the foramen ovale must
have greatly aggravated the congestion of the systemic
venous system.[1]

PERMANENT PATENCY OF THE FŒTAL PASSAGES.

Allusion has before been made to the persistence of the
fœtal passages when combined with imperfect separation of
the ventricular cavities.   It is not, however, only when the
septum of the ventricles is deficient that the foramen ovale
and ductus arteriosus continue pervious; but under all circumstances the persistence of these passages is very generally
associated with some obstruction at or near the pulmonic
orifice.

Patent foramen ovale.

When the foramen ovale remains open, it may be under
the following circumstances :—

1st. The opening may be unusually large, and the valve
entirely wanting.

2ndly. The foramen may be larger than usual or of the
natural size, but the folds may be imperfectly developed so
as to be too small to close the opening.[2]

3rdly. The opening may be natural and the valve may

---

[1] Path. Trans., vol. xi. 1859–60, p. 40.   Preparation in Victoria Park Hospital Museum, B 19.

[2] This defect is shown in plate 7, fig. 4, from a preparation in the Museum of
the Victoria Park Hospital, numbered B 10.

3rdly. The opening may be natural and the valve may have attained the full size; but the membrane may be very defective, being perforated by one or more larger or smaller sized apertures.

4thly. The foramen may be of natural size, and the valve may be fully developed, so as to be capable not only of covering the orifice but of reaching some distance above its upper edge; but the changes which should occur shortly before and after birth may not have taken place. The cornua of the valve may retain the length which they have at the later periods of fœtal life, and may continue widely apart, so that the fold may hang loosely across the orifice and offer but little impediment to the continued passage of the blood from the right into the left auricle.[1]

5thly. The valve may completely cover the opening so as under ordinary circumstances entirely to obstruct the flow of blood through it, yet the fold may not become adherent to the edges of the aperture, and an oblique passage may still remain.

Of these irregularities, the three first are rarely seen except when the heart presents other serious defects in its conformation. The latter condition, on the contrary, is of very common occurrence; and is found in cases in which, in an organ otherwise well formed, there has been some source of obstruction to the flow of the blood from the right auricle or ventricle at the time of birth.[2]

The closure of the foramen ovale is generally ascribed to the floating up of the valve above the margin of the opening; and to the fold being retained in apposition with the isthmus, by the pressure of the blood which enters the left auricle after the establishment of the pulmonary circulation. On this theory the non-closure of the

---

[1] This condition is shown in plate 3, fig. 3, from the specimen described in Case xiii. page 112. The preparation is contained in the Museum of the Victoria Park Hospital, and is marked B 3.

[2] See also Observations by Mr. Struthers, in the Edin. Jour. of Med. Sc., vol. xv. 1852 (3rd series, vol. vi.), p. 21. Republished in Anatomical and Physiological Observations, part 1. Edinburgh, 1854, p. 63.

orifice has been referred to the undue distension of the right auricle, owing to obstruction to the flow of blood from that cavity, causing the valve to be pushed aside from the septum and preventing the adhesion of the membranes. It is, however, very questionable whether the process by which the opening becomes closed is so purely passive as is thus supposed. On examining the heart in young children at different periods after birth, it will be found that the obliteration of the passage is effected—

1st. By the shortening of the cornua, and the drawing up of the fold of the valve considerably above the edge of the foramen.

2ndly. By the approximation of the cornua, so that the width of the upper edge of the valve is greatly reduced; and,

3rdly. By the diminution of the opening itself.

For the production of these changes special provisions exist:—In the hearts of children a few weeks or months old, muscular fibres can readily be traced proceeding from the fasciculi which form the walls of the left auricle, in the course of the cornua of the valve, to be expanded on the curtain. Of these fibres, those from each cornu go to opposite sides; while others, as described by Senac, pass across the fold, following a course more or less parallel to the upper edge of the valve. It is evident that muscular fibres thus placed, must, when they contract, tend, on the one hand, to draw up the curtain so as to cause it to overlap the edges of the opening; and on the other, to approximate the cornua. The opening itself is bounded by the muscular fasciculi which form the inter-auricular septum, and their contraction must reduce the size of the orifice, as also pointed out by Senac. There is therefore every reason to regard the closure of the foramen as an active process, dependent on the contraction of these different sets of muscular fibres; induced, probably, partly by the traction exercised upon them by the expansion of the cavity of the left auricle, from the increased quantity of blood which enters it after birth; and partly by the more stimulating quality of the blood which reaches the cavity or circulates through the cardiac vessels.[1]

If this explanation of the mode of closure of the foramen ovale be correct, the persistence of the opening in some cases of malformation may not be entirely caused by the undue distension of the right auricle, but may partly depend on the quantity of blood entering the left auricle being insufficient to exercise the due amount of traction, or too imperfectly acted upon by the air to produce the requisite excitement. Cases occasionally occur in which the foramen is found closed, though there must always have existed much greater pressure on the right than on the left side, and such cases do not admit of explanation on the ordinarily received theory. The continuance of undue pressure on the right side of the valve for a considerable period after birth, owing to the imperfect establishment of the pulmonary circulation, may, however, prevent the adhesion of the valve to the margin of the opening, and so cause the persistence of the oblique passages, which, as before mentioned, frequently exist between the cavities.

When the foramen ovale is still pervious, the size of the opening varies in different cases. It may possess a diameter of half an inch to an inch, or it may be of much smaller size, only consisting of a narrow fissure between the upper edge of the valve and the margin of the foramen. When it is largely open, the Eustachian valve often still exists and is fully developed. There is also very generally some obstruction at the pulmonic orifice and this is sometimes extreme. I have already mentioned a case which occurred to M. Bertin, in a female fifty-seven years of age, in which the orifice of the pulmonary artery was reduced by disease of the sigmoid valves, to a passage two and a half lines in diameter (5·62 mm. ·222 E. in.) ; and other scarcely less remarkable instances of contraction might be quoted. The alteration in

---

[1] See remarks on the mode of closure of the foramen ovale, by the author, published in the Path. Trans., vol. iv. 1852-53, p. 85. The preparations which illustrate these remarks are contained in the Museum of the Hospital for Diseases of the Chest, Victoria Park.

the form of the heart which ensues in consequence of this condition is very marked, and corresponds with that which has been before referred to as resulting from the pulmonic obstruction, when the septum of the ventricles is imperfect.[1] There is, however, some difference in these changes, according to the degree of obstruction and the freedom of communication between the auricles. In some instances, where the contraction at the pulmonic orifice is great and the foramen ovale largely open, the right ventricle is found small and its walls flaccid, and the hypertrophy and dilatation are chiefly manifested on the left side;[2] and this is especially the case when the right auriculo-ventricular aperture also is contracted.[3]

1. *Open foramen ovale, with constriction of the pulmonary orifice.*

A case of the description of malformation now referred to was related by Morgagni in 1761.[4] It occurred in a girl fifteen years of age, who had been ailing throughout her life, and suffered from shortness of breath and debility, and was generally livid. The heart was small and rounded at the apex; the right auricle was much enlarged and its walls thickened, and the foramen ovale was so imperfectly closed that the little finger could be passed through the opening. The orifice of the pulmonary artery was greatly contracted owing to disease of the valves. They were cartilaginous and in places osseous, and were so connected together as to leave only an aperture sufficient to admit a barleycorn. In 1783, Tacconi[5] related an instance of the same kind of defect to the Academy of Sciences of Bologna, which he met with in a girl

---

[1] See page 38.
[2] The case of Dr. Hallowell, Am. Jour. of Med. Sc., vol. xxii. 1838, p. 366, affords an example in point.
[3] This occurred in the case of M. Bertin.
[4] De Sed. et Causis Morb., Venetiis, 1761, tomus primus, f. 154, and Alexander's translation, 1769, vol. i. letter 17, arts. 12 and 13, p. 435. Referred to elsewhere as Valsalva's case.
[5] Comment. Bonon., 1783, p. 64.

fifteen years of age, who had been livid, incapable of active exertion, susceptible to cold, and subject to dyspnœa, cough and expectoration, mixed occasionally with blood. In 1805, a similar malformation was found by Seiler,[1] in a man twenty-nine years of age; in 1810, by Schuler,[2] in an infant ten weeks old; and in 1817, by Poliniere,[3] in a boy of fifteen. In the latter case the symptoms did not appear till after eight years of age; the boy then began to manifest the usual signs of malformation, and he had repeated attacks of bronchitis and hæmoptysis, during which he became very livid. The right auriculo-ventricular aperture was also found contracted. Since the last date cases of a similar kind have been reported by Cherrier,[4] Bonnissent, Obet and Pinel;[5] Bertin and Lallemand,[6] Hallowell,[7] Lombard,[8] Urban,[9] Craigie,[10] Spitta,[11] Leared,[12] Struthers,[13]

---

[1] Horne's Archiv, 1805 ; Hein, obs. 29.

[2] Dissert. de Morbo-Cœruleo, Œniponte, 1810 ; Hein, obs. 35.

[3] Bibliothèque Médicale, t. lvii. 1817; Gintrac, obs. 44.

[4] Thèse, 1820, No. 252, p. 24 ; Gintrac, obs. 48, in a male thirty-four years of age.

[5] Rev. Méd., t. vi. 1821, p. 175 ; Gintrac, obs. 50me, in a female five years of age.

[6] Recherches sur l'Encephale, 1825, t. ii. p. 7 ; Gintrac, obs. 52, in a female fifty-one years of age, with contraction of the right auriculo-ventricular aperture.

[7] American Jour. of Med. Sc., vol. xxii. 1838, p. 366, in a female child six months old.

[8] Mem. de la Soc. de Phys. et d'Hist. Nat. de Génève, t. viii. 1839, p. 503, in a female twenty-eight years old.

[9] Tiedemann's Verengung etc. der Pulsädern, p. 116. Pulmonary artery so contracted as only to admit a small quill. In a female eleven and a half years of age, who had been cyanotic.

[10] Edin. Med. and Surg. Jour., vol. lx. 1843, p. 271, in a male twenty years of age.

[11] Med.-Chir. Trans., vol. xxix. 1846, p. 81, in a female forty years of age.

[12] Dublin Journal of Medical Science, N. S., vol. x. 1850, p. 223. The subject was a boy eight years of age. The foramen ovale was unclosed over a small space, and the valve itself was defective.

[13] Edin. Monthly Jour. of Med. Sc., 1852, 3rd series, vol. vi. p. 21 ; and Anatomical and Physiological Observations, part 1. Edinburgh, 1854, p. 63. In a child which lived fifteen months. In this case it is stated that the ductus arteriosus was not obliterated, but was incapable of transmitting blood.

Gordon,[1] Speer,[2] M. Perati,[3] Dr. Wilks,[4] and Dr. Andrew,[5] and the following case which occurred in my own practice, was related at the Pathological Society in 1847.[6]

CASE XIII.—*Contraction of the orifice of the pulmonary artery; communication between the cavities of the auricles by the foramen ovale; septum of the ventricles entire.*

A young man, twenty years of age, by trade a pipe-maker, was admitted into the Royal Free Hospital under my care, on the 20th May, 1847. He stated that he had enjoyed good health till two years before. Since that time he had been declining in strength, and had suffered from cough, difficulty of breathing, palpitation and pain in the left side of the chest. He was occasionally observed to be somewhat livid in the face, and had expectorated a considerable quantity of blood six weeks before. He had never had rheumatic fever.

When admitted into the hospital he was delicate-looking, and emaciated; his face pale, and the expression of countenance anxious. The lips were slightly livid, the fingers long and a little clubbed at their extremities, and the nails incurvated and somewhat dark-coloured. He had urgent dyspnœa and a severe cough, and expectorated a large

---

[1] Trans. of Path. Soc. of Dublin, referred to by Dr. Stokes, Diseases of the Heart and Aorta, 1854, p. 166. In a boy twelve years of age.

[2] Med. Times and Gaz., N. S., vol. xi. 1855, p. 412. In a female seventeen years of age.

[3] Bulletin de la Soc. Anat., 2mo série, t. iii. 1858, p. 450. In a cyanotic child who died at the age of six years.

[4] Path. Trans., vol. x. 1858–59. In a female eighteen years of age.

[5] Path. Trans., vol. xvi. 1864–65, p. 81. In a female, a patient of Dr. Edwards, aged six years, who was slightly dusky, but not decidedly cyanotic. The orifice of the pulmonary artery was very greatly contracted from fusion of the valves, and the vessel itself was small and its coats thin. The aortic aperture had only two valves—one of them displaying union of two segments. The tricuspid valves were thickened and contracted. The foramen ovale widely open.

[6] Path. Trans., vol. i. 1846–47, 1847–48, p. 200. The specimen is preserved in the Victoria Park Museum, and it is numbered B 3. It is engraved in plate 3, figs. 1, 2, and 3.

quantity of muco-purulent fluid. The pulse was quick and feeble. The sternum was prominent, and its lower end was directed to the left side so as to occasion a decided projection of the third, fourth, fifth, and sixth left cartilages, and a corresponding flattening of those of the opposite side. Beneath the right clavicle the chest was contracted and imperfectly resonant, and on the left side it was fuller and more sonorous. The respiratory sounds were masked by subcrepitant and mucous râles. The liver was considerably enlarged. The præcordial dulness commenced above at the level of the third left cartilage, and extended transversely from the right side of the sternum to beyond the line of the left nipple. The impulse of the heart was felt over a large portion of the front of the chest, and was attended by a loud murmur audible over the whole dull space. This murmur was most intense below and within the line of the left nipple, and there the second sound was entirely masked by it. It was also heard very loudly between the sternum and nipple, and thence towards the middle of the left clavicle. At the top of the sternum it was less distinct and of shorter duration, so that it was followed by a clear second sound. To the right of the lower part of the sternum, there was heard first an imperfect systolic sound, and this was followed by, or lapsed into, a short murmur, which was again succeeded by a clear diastolic sound. A purring tremor was felt in the præcordia. There was no anasarca.

During the time the boy was in the hospital, he continued to sink gradually and died eleven days after admission.

On examination of the body, the brain was found healthy, but the skull afforded a very marked instance of the unequal growth of the two sides often observed in rickety children.

The right lung was extensively permeated by tubercle, and towards the apex contained several small cavities. There was much solid tubercle in the left lung. The bronchial mucous membrane was reddened and the tubes contained much muco-purulent fluid.

I

There was a considerable amount of serum in the cavity of the pericardium, and the attached membrane was in several places covered by rough and thick patches of old lymph. The heart was situated more towards the left side than usual, and the right ventricle was turned forwards, so that it constituted the whole exposed part of the organ ; while the left ventricle was situated entirely posteriorly.

The heart was of large size, and weighed 12 oz. avoirdupois.[1] The valves of the pulmonary artery were adherent together so as to form a complete diaphragm extended across the opening into the vessel, and perforated in the centre by an irregular triangular aperture of sufficient size to admit the passage of a lead pencil. This membrane was protruded forwards so as to project into the cavity of the vessel, and displayed three ridges marking the former points of separation of the valves, with deep sacs between each. Around the aperture there was a ring of small vegetations. The right auricle and ventricle were greatly dilated, and their walls hypertrophied. The parietes of the ventricle measured near the pulmonic orifice, four French lines (9· mm. ·35 E. in.) in width. The muscular substance was unusually firm. The right auriculo-ventricular valves were natural. The foramen ovale was so widely open as to allow of the passage of a ball measuring three French lines in circumference (6·75 mm. ·26 E. in.). The Eustachian valve was of large size. The left auricle and ventricle were, relatively to the right cavities, small in capacity, and the walls of the latter were of moderate thickness and flaccid. The aortic and mitral apertures and valves were natural. The ductus arteriosus was impervious. The liver, spleen, and kidneys, were of large size. The glands of Peyer in the ileum displayed small tuberculous depositions, and in places commencing ulceration.

2. *Open foramen ovale without other defect.*

In the malformations of which the cases now related and referred to afford examples, the foramen ovale appears to

---

[1] The weight should not have exceeded 8½ or 9 oz.

remain open in consequence of the obstruction at the pulmonic orifice. Instances are, however, occasionally met with, in which with an open state of the foramen ovale, the pulmonary artery and orifice are free from disease, and there is no source of obstruction to which the defect can be ascribed. A case of this kind, which has frequently been quoted, was related by Dr. Spry, in 1805.[1] A similar case has also been placed on record by Dr. Mayo, in which the foramen ovale was found largely open in a female who died at the age of seven years, and had never presented during life any appearance of cyanosis.[2] The foramen ovale was so largely open as to be two inches in circumference, and the ductus arteriosus was also pervious. There appears to have been no obstruction in the pulmonary artery or at the outlet of the right ventricle. The subject of the case was a cyanotic female, seventeen years of age. In the Museum of St. Thomas's Hospital there is also a preparation illustrating the same condition. It is marked LL 61, and is stated to have been removed from a female, twenty-one years of age, who, since she was three months old, had presented characteristic symptoms of malformation of the heart, cyanosis, palpitation, dyspnœa, faintings, occasional convulsive attacks and lividity. The foramen ovale is so largely open as to have a diameter of fully one inch, and the valve is entirely wanting; the heart being otherwise free from disease, except some thickening of the mitral valve. Another preparation, numbered LL 60, exhibits a largely open foramen ovale without any rudiment of the valve, in the heart of an infant. I have also given a drawing of an imperfectly closed foramen ovale, from a preparation which exists in the collection at the Hospital for Diseases of the Chest, Victoria Park.[3] It was removed from a girl, eight years of age, who was under my care at that institution. She was reported to have been always

---

[1] Memoirs of the Medical Society of London, vol. vi. 1805, p. 137.

[2] Lond. Med. Gaz., vol. i. N. S., 1843, p. 613.

[3] Victoria Park Museum, B 16. This specimen is described in the Path. Trans., vol. ii. 1848–49, 1849–50, p. 183.

delicate but to have had no serious indisposition till she took measles, two years before her death. She died with the usual symptoms of cardiac asthma, and there was no decided cyanosis at any period of her illness. On examination, the valve of the foramen ovale was found very imperfect, so that an aperture remained sufficiently large to admit the passage of the fore-finger. The aortic orifice and aorta were small. The pulmonary orifice and artery were, on the contrary, much larger than usual.

In this case it is doubtful what may have been the cause of the imperfect closure of the foramen. It is possible that the small size of the aortic orifice may have occasioned it ; but it is much more probable that the smallness of the aorta resulted from some obstruction in the pulmonary branches or lungs, which, though showing its effects in the dilated pulmonary artery and open foramen ovale, had itself disappeared. The imperfect closure of the foramen ovale may, indeed, depend on a variety of causes. The most frequent are, doubtless, obstruction at the pulmonic orifice, in the right ventricle, or at the right auriculo-ventricular aperture ; but impediments to the flow of blood through the lungs, or obstruction at the left auriculo-ventricular aperture, or at the orifice of the aorta, would probably equally prevent the obliteration of the passage.

The following case affords a still more remarkable example of this condition unconnected with any other defect in the conformation of the heart :—

CASE XIV.—*Largely open foramen ovale, without other defect in the heart.*

The patient was a female, aged sixteen, who died in St. Thomas's Hospital, in March, 1860. She was first taken ill in September, 1858, while residing at Luton, in Bedfordshire. She came up to town, and was so ill for the first six months of 1859 as to be incapable of leaving the house or following her occupation—that of the straw-plait.

She suffered, at the time, from catamenial irregularity and symptoms of general debility. In the summer she recovered so as to be able to resume her occupation, and continued better till the beginning of the former year. She applied at the Victoria Park Hospital in February, and was admitted into St. Thomas's on the 16th of March. While under observation she was obviously sinking rapidly under what appeared to be acute miliary tubercle. She had great difficulty of breathing, cough, and expectoration, and stated that she had occasionally spat a little blood. There was general deficiency of resonance on percussion at the chest, especially the upper parts, with the usual signs of bronchitis; and she was emaciated and much prostrated. She died on the 22nd of March.

On inquiry after death, it appeared that she had always been delicate, having had fits in early life. After six years of age the fits ceased and she became stronger; but, though possessed of considerable intelligence, she could not be taught to read from the pain in the head occasioned by mental exertion. Her father died of some cerebral affection when seventy-two years of age, and the three other children of her parents all died in early life. There were never any symptoms of cyanosis, nor had anything at all peculiar in her appearance attracted the notice of her mother; but it was mentioned, that during her last illness she suffered occasionally from palpitation, especially on active exertion; and when hurried, as by ascending stairs, she became breathless, faint, and somewhat dark in the face.

On post-mortem examination, the left pleura was found firmly adherent above, and the lung, in the rest of its extent, was invested by a film of false membrane. There were also adhesions on the right side, which were more extensive and of older date than those on the left.

Both lungs were the seat of crude tubercular deposit, more abundant above than below, and more advanced on the right than on the left side. The tubercles in many places were breaking down; and at the left apex there were two or three cavities, each about the size of a chest-nut. The intervening lung tissue, especially in the upper lobes, was nearly solid and airless, and infiltrated by inflammatory deposit. The bronchial tubes were full of muco-purulent secretion. The larynx and trachea were healthy.

The peritoneal surface was covered by fibrous tissue, the result of chronic inflammation, which produced firm adhesions between the liver and diaphragm, and covered the small intestines with irregular flocculi. On closer examination, the surface was found studded thickly with grey miliary tubercles, which were most abundant in the lumbar regions and over the intestines. Connected with the surface of the mesentery and between the liver and diaphragm, there were, in addition to the miliary deposits, a few masses of crude tubercle, from the size of a horse-bean downwards. The spleen was rather large ; the other organs healthy.

The heart, when removed from the body, weighed seven ounces and three-quarters avoirdupois. The right ventricle was considerably dilated and the walls somewhat thicker than usual. The left ventricle was of natural capacity, and its walls retained their normal width. The tricuspid valve, with the right auriculo-ventricular aperture and the pulmonic and aortic apertures and valves, were natural; but the left auriculo-ventricular aperture was small (giving passage only to a ball measuring thirty-six French lines (81 mm. 3·19 E. in.) in circumference), and the valves were white, thickened, and somewhat rigid. The auricles were both of large size. The foramen ovale was entirely unclosed; indeed, there were merely the rudiments of the valve, and the aperture freely admitted the passage of a shilling.[1]

It is well known, as already mentioned, that the valve of the foramen ovale does not always become entirely adherent to the edges of the fossa, so that a space is left which admits of the passage of some small or flat body, as a probe or the handle of a scalpel, from the one auricle into the

---

[1] Path. Trans., vol. xi. 1859-60, p. 68. Preparation in Victoria Park Hospital Museum, B 20.

other. Under ordinary circumstances the obliquity of this passage entirely prevents the flow of blood along it; but it has been thought that when the right auricle is greatly distended and the fossa much enlarged, the fold not becoming proportionately expanded, may cease to overlap the upper edge and the passage be restored. Laennec contended that "a blow, fall, or violent exertion might cause the dilatation of the oblique opening;"[1] and a case related by Corvisart[2] has been referred to as an example of an open condition of the foramen ovale so produced. A postilion, fifty years of age, after having received violent blows on the epigastrium, suffered from symptoms of cardiac disease, with great dyspnœa and lividity; and after death the heart was found much enlarged, and an opening existed between the auricles capable of admitting the points of four fingers. I possess a preparation,[3] removed from a female aged fifty-four, who died of old and extensive disease of the lungs combined with bronchitis, in which the membrane closing the oval opening is expanded into a sac of considerable size, which projects into the left auricle. The valve is not entirely adherent to the margin of the fossa, so that an oblique passage from the right to the left auricle exists. The fold, however, still overlaps the edge of the opening, and the passage of blood between the two cavities must have been effectually prevented. This specimen therefore shows, that though the valve be not entirely adherent, it may yield to pressure so as to be expanded, rather than allow of being drawn down; and it is probable that, when a communication occurs in after life between the two auricles, it is rather due to the rupture or erosion of the valve, than to the restoration of the passage. In Corvisart's case the opening appears to have been produced by the destruction of a portion of the valve; and Bouillaud strongly maintains that, in some cases, the communications which take place between the auricles are the

---

[1] Diseases of the Chest, Forbes' translation, 4th edition, 1834, p. 575.
[2] Essai sur les Maladies du Cœur, 3me éd., 1818, p. 290, obs. 44.
[3] Victoria Park Hospital Museum, C 20.

result of disease. In a preparation of partial aneurism of the heart which I possess,[1] the valve of the foramen ovale is perforated by a large aperture, probably produced by endo-carditis and erosion.

Since, however, the first edition of this work appeared, I have met with a case which shows that Laennec's views may in some instances be correct. It occurred in a female, twenty-four years of age, who had curvature of the spine and deformed chest, but was reported to have been in usual health till a month before her death. She laboured during her illness under great dyspnœa, headache, transient deli-rium, irregularity of the pulse, coldness and very marked lividity of the surface, and died comatose. The heart was large, and weighed $12\frac{1}{2}$ oz. avoirdupois. There was great dilatation of the right auricle, ventricle, and pulmonary artery; and the walls of the right ventricle were much hyper-trophied, measuring $4\frac{1}{2}$ lines (10·2 mm. ·39 E. in.) in width. The fossa of the foramen ovale was very greatly expanded, and the valve did not entirely cover it—an aperture existing between its upper edge and the isthmus three lines in width.[2]

There can seldom, I think, be any difficulty in distin-guishing cases of reopening of the foramen ovale, the result of disease in after-life, from those in which the defect in the separation of the auricles is of congenital origin.

3. *Congenital contraction of the pulmonary orifice with occlusion of the foramen ovale.*

It has been shown that, though the patency of the fora-men ovale generally coexists with some form of obstruction in the right ventricle or pulmonary artery, such is not always the case. It occasionally also occurs that even marked obstruction at the pulmonic orifice is found when

---

[1] Victoria Park Hospital Museum, C 23; Edin. Med. and Surg. Jour., vol. lxvi. 1846, Case I. p. 263.

[2] Path. Trans., vol. x. 1858–59, p. 108. This preparation is in the Victoria Park Hospital Museum. A case probably of similar character is described by M. Durozier in the Comptes Rendus de la Soc. de Biologie, 1862, p. 105.

the foramen ovale is entirely impervious, and the heart is in every other respect natural. Instances of this kind have been recorded by Philouze,[1] Chelius,[2] Burnet,[3] Fallot,[4] Craigie and Graham,[5] Tiedemann and Fohman,[6] Cruveilhier,[7] Ogle,[8] and Dr. Wale Hicks,[9] and a specimen affording a remarkable example of the same condition was exhibited at the Pathological Society in 1852, by Dr. Hamilton Roe. In reference to several of these cases it may, and has indeed, been contended, that the disease of the pulmonic valves was not congenital. From, however, the precise similarity of the condition of the valves in cases of this description to that which is found in instances of undoubted malformation, I am disposed to believe that the union of the valves takes place during intra-uterine life. This inference is confirmed by the fact that in some of these cases the valves are found thus united in children that die at very early periods of life, and without having previously displayed any evidence of cardiac

---

[1] Bulletin de la Soc. Anat. de Paris, t. i. 1re année, 1826, p. 158. The age of this patient is not given.

[2] Quoted by Tiedemann in Verengung, etc., from the Heidelberger Klinischen Annalen, 1827, in a man twenty-six years of age, who served as a soldier, and was healthy till he was suddenly seized a year before his death. He laboured under the usual symptoms of heart disease with lividity. The pulmonary artery was reduced from adhesion of the valves to a mere chink, one line broad and three lines long.

[3] Journal Univ. Hebd. de Méd. et de Chir., 1830, t. i. 1re année, quoted by Bouillaud, Traité, etc. 2me éd. t. ii. p. 281, obs. 28. In a female seven years of age, who had been ill six months, and was livid, and the extremities œdematous. The case of M. Jadelot.

[4] Quoted from a Belgian Journal in Lond. Med. and Surg. Jour., vol. v. 1834, p. 61, in a female sixty-three years of age.

[5] Edin. Med. and Surg. Jour., vol. lx. 1843, p. 277, in a man forty-four years of age.

[6] Op. cit., p. 20. Referred to in analysis of Tiedemann on Arctation and Closure of Arteries, in the Edin. Med. and Surg. Jour., vol. lxv. 1846, p. 127, in a man twenty-one years of age.

[7] Anat. Path., liv. xxviii. plate 4, fig. 2.

[8] Path. Trans., vol. v. 1853–54, p. 69, in a female fourteen years of age.

The specimen drawn by Dr. Carswell, in his Illustrations of the Elementary Forms of Disease—Hypertrophy—probably offers another example of the same condition occurring during fœtal life. It was taken from a female above forty years of age, but full particulars of the case are not given.

[9] In a child two years and three-quarters old, who died of croup, and was said to have never been previously ill.—Path. Trans., vol. xvi. p. 91.

or other disease.[1] There can, however, be little doubt that when
the symptoms do not manifest themselves till late in life,
the defect, if congenital, must have been originally slight
and have been greatly aggravated by subsequent disease.
Such an extreme degree of contraction of the pulmonic
orifice as occurred in some of these cases, could certainly
not have existed at the period of birth without preventing
the occlusion of the foramen ovale.

The following case affords an example of this kind :—

CASE XV.—*Congenital contraction of the orifice of the pulmonary artery from fusion of the valves ; no other defect in
the heart and the foramen ovale closed.*

The subject of this defect was a man, aged twenty-three,
who first came under my care in November, 1855. He
stated that he had then been ill eighteen months, and that
his indisposition commenced with profuse spitting of blood.
He had subsequently had two other attacks of hæmoptysis,
and suffered from cough and expectoration ; his voice was
very husky ; the percussion sound under the left clavicle
was defective, and the respiratory sounds were feeble in that
situation, the cough and vocal resonance being increased.
A loud systolic murmur was audible along the upper part of
the sternum, and especially in its left side.

In the spring of 1856 he was admitted into the Victoria
Park Hospital, and again became a patient at that institution in January, 1859. In the interval he attended as an
out-patient, and during that period his phthisical symptoms
gradually advanced. When last admitted he was incapable
of speaking except in a whisper ; he had an abortive but
frequent cough, and profuse expectoration, and the usual
symptoms and signs of advanced consumption. The cardiac
murmur continued as before. He was suddenly seized with
cerebral symptoms, and died comatose in February.

---

[1] Dr. Hicks' case.

Throughout his illness he never presented any appearance of cyanosis, and he had a peculiarly clear and pallid complexion.

On post-mortem examination, the brain was found much congested. The larynx was extensively ulcerated, and the chordæ vocales entirely destroyed. The lungs contained interspersed tubercles and cavities, especially on the left side. The heart was of its natural size, and weighed nine ounces and three-quarters avoirdupois. There was great hypertrophy of the walls of the right ventricle, which in front measured four lines in thickness (9· mm. ·35 E. in.) The pulmonic orifice was reduced to about half its normal capacity, the contraction being due to the adhesion of the valves. These had originally been three in number, but they were united and protruded forwards in the course of the pulmonary artery and much thickened. The aperture only allowed of the passage of a ball measuring fifteen French lines in circumference (33·75 mm. 1·33 E. in.), while the trunk of the artery itself admitted a ball measuring thirty-three lines (74·25 mm. 2·93 E. in.), and the aortic orifice, one measuring thirty-nine lines (87·75 mm. 3·46 E. in.). The other valves were healthy : the foramen ovale was entirely closed, and the heart otherwise well formed.[1]

4. *Patent Ductus Arteriosus.*

It has been already mentioned that when the orifice or trunk of the pulmonary artery is obliterated during foetal life, the ductus arteriosus very generally remains open and forms the means by which the blood is conveyed to the lungs. When also there is only slight contraction of the pulmonic orifice the passage not unfrequently continues pervious. The very general persistence of the arterial duct, when the origins of the main arteries are transposed and the commencement of the descending aorta is contracted or obliterated, will be referred to hereafter. At present it is proposed only to

---

[1] Path. Trans., vol. x. 1858–59, p. 107. Preparation B 18, Victoria Park Hospital Museum.

speak of the open state of the ductus arteriosus when that
condition is found in hearts otherwise well formed; and the
closure of the passage is prevented by the existence of some
obstruction to the circulation of the blood at the time of
birth.

Under ordinary circumstances, the closure of the duct is
effected by a process of general contraction, which commences at the aortic extremity and gradually advances
towards the bifurcation of the pulmonary artery. It is
completed at the aortic end while the duct remains pervious
at the pulmonic extremity, and thus presents a fusiform
cavity, having its apex near the aorta and its base opening
into the pulmonary artery. In some instances, however,
when the blood has been partially supplied to the pulmonary branches for some time after birth through the
ductus arteriosus, the mode of closure is reversed; the contraction commencing at the pulmonic extremity and gradually advancing towards the aortic, so that there sometimes
exists a fusiform cavity, having its apex at the bifurcation
of the pulmonary branches and its base in the aorta. These
partially pervious states of the duct are occasionally seen in
cases of malformation of the heart where the process of
closure has been delayed.

When the duct remains open it may be found of larger
size than is usual at the time of birth, as in the cases
before mentioned in which the foramen ovale was obliterated during intra-uterine life. It may retain its natural
size, as when the aorta beyond the left subclavian artery is
greatly obstructed or impervious. It may be reduced to
a narrow canal only capable of admitting a probe, a crow-
quill, or a small goose-quill. Or, lastly, it may be found
to present a sacculated dilatation, as in a case mentioned
by Billard,[1] and of which examples appear to have occurred
to Baron,[2] Martin,[3] Parise,[4] Chevers,[5] and Thore. The last
author has, indeed, made this condition the subject of a

---

[1] & [2] Maladies des Enfans nouveaux nés, Paris, 1828, p. 567, with a plate.
[3] Bullet. de la Soc. Anat., 2mo année, 1827, p. 17.
[4] Ibid., 12mo année, 1837, p. 95.   [5] Path. Trans., vol. i.

special memoir, in which he describes eight cases of the kind in infants which survived for various periods from four days to a month.[1]

The length and form of the duct also may vary :—It may retain its natural length, or be considerably shorter than usual, or may be represented only by a direct opening between the pulmonary artery and the arch of the aorta, as in the case reported by Lediberder.[2] Its position also may be irregular ; the most common deviation being, that instead of proceeding from the point of bifurcation of the pulmonary artery to the descending aorta below the origin of the left subclavian artery, it arises from the left branch of the pulmonary artery, and is inserted into the aorta opposite the left subclavian artery ; or into the left subclavian artery itself;[3] or at a still higher point of the aorta, as under the arch.   Indeed it would appear that the under portion of the aortic arch does not become developed in proportion to the growth of the upper part, from which the vessels to the head and upper extremities are given off; and thus in the adult, the cord of the obliterated duct is often found attached to the under part of the aortic arch ; instead of to the descending aorta below the left subclavian artery.   In yet other cases the distribution of the duct may be irregular, so that it may give off the left subclavian artery, as in the cases recorded by Reinmann,[4] Holst,[5] and Hildenbrand ;[6] or, as in one before quoted from Breschet, there may be two ducts, one proceeding from the left pulmonary branch to the aorta, the other from the right pulmonary artery to the brachio-cephalic trunk.   In a case observed by Wrisberg,[7] the duct arose as a distinct vessel from the right ventricle.

---

[1] Arch. Gén. de Méd., 4ᵐᵉ série, t. xxiii. 1850, p. 30.

[2] For reference, see p. 75.

[3] This was the case in the instance, figured and described by Sandifort (Museum Anatomicum, vol. i. f. 273 ; and vol. ii. f. 107, figs. 1 and 2), and also by Agliette (Saggi di Padova, t. i. 1786, p. 69), in the cases of Calliot and Obet, and in that of Jackson, Med. and Phys. Jour., vol. xxxiv. July to Nov. 1815, p. 100.

[4] Nova Arch. Physico-Medica, 1757, No. 12, obs. 74, p. 302.

[5] P. 58.   [6] P. 58.   [7] Hein, obs. 67.

In cases like those last alluded to, the patency of the duct is combined with the irregular distribution of the vessels from the faulty development of the branchial arches; but more commonly the vessels are regularly developed, and the duct remains open in consequence of some obstruction to the transmission of the blood through the lungs or systemic vessels; and, as these causes would also determine the persistence of the foramen ovale, the patency of the two passages is not unfrequently coincident.

I have already related two cases in which, with an imperfect state of the inter-ventricular septum, the duct remained unclosed, and the observations recorded by Obet, Haase, Cheevers, Graves, and Houston, may be referred to as furnishing examples of the same condition.[1] In the cases of Seiler and Schuler, the septum of the ventricles was entire, but the foramen and duct were both open. In all these cases the patency of the duct was due to the existence of obstruction to the exit of the blood from the right ventricle at the time of birth. It is probable, also, that contraction of the right auriculo-ventricular aperture and imperfect expansion of the lungs, by preventing the free circulation of the blood after birth, would equally determine the patency of the duct. A case mentioned by Richerand,[2] in which the septum of the ventricles was imperfect and the duct unclosed with the pulmonary artery very wide, in a person forty-one years of age, may have depended on the former existence of obstruction in the pulmonic circulation. Contraction of the left auriculo-ventricular aperture and of the aortic orifice may also prevent the closure of the duct. A case is related in the Transactions of the Pathological Society of Dublin, by Dr. Mayne,[3] in which the foramen ovale and duct were open, probably from disease of the left

---

[1] Several cases of persistence of the ductus arteriosus are given by Rokitansky in his work, Ueber einige der wichtigsten Krankheiten der Arterien. Wien, 1852. Beob. 14, 15, 16, 17, and 18. And he also quotes a case from Skoda, Beob. 13, Taf. 13. These cases were in persons thirty-two, twenty-two, twenty-one, forty-three, and twenty-three years of age.

[2] Él. de Physiologie, 1811, t. i. p. 205.

[3] Transactions, 1847 to 1852, p. 35. Also published in the Dublin Quarterly Journal of Medical Science, vol. v. No. 9, N. S., 1841, p. 46.

auriculo-ventricular aperture, which was so contracted as only to admit two fingers, while the pulmonary orifice admitted three and the right auriculo-ventricular aperture four fingers. The subject of the affection was a female, twenty-seven years of age, who suffered from dyspnœa and palpitation with a short dry cough, but did not display the slightest cyanosis. The foramen ovale was sufficiently large to allow a half-crown to pass through, and was apparently unprovided with any valve.

In the first volume of the Pathological Transactions,[1] a case is related by Dr. Barlow, which occurred in the practice of Dr. Babington, and is probably an example of an open ductus arteriosus dependent on aortic obstruction. The aortic orifice was of small size, and the valves were four in number. They were extensively diseased, and not only caused obstruction to the flow of blood from the ventricle but were incompetent to prevent its return. There was a contraction of the aorta behind the left subclavian artery, and below this point an opening as large as a goose-quill, formed a direct communication between the aorta and pulmonary artery. The patient, a female thirty-four years of age, was born at the seventh month, and suffered from symptoms of cardiac disease from early life. She had never had acute rheumatism. A case related by Dr. Hare, in the Pathological Transactions, affords a still more decided example of obstruction at the aortic orifice occasioning an open state of the arterial duct. The child died with cardiac symptoms at the age of six months, and the aortic orifice was found of small size, and only provided with two valves; there was a large deficiency in the septum of the ventricles, and the pulmonary artery was of large size.[2]

In the following case the duct apparently remained open in consequence of the existence of some degree of contraction in the aorta distal to the left subclavian artery or above the point of insertion of the duct :—

---

[1] P. 55.   [2] Path. Trans., vol. xi. 1859–60, p. 46.

CASE XVI.—*Open ductus arteriosus with narrowing of the commencement of the descending aorta.*

The preparation was removed from the body of a man, aged thirty, who was admitted into St. Thomas's Hospital, on the 6th of February, 1860, but who died in the bath before he could be placed in bed. No history of his previous condition could be obtained; but, while under observation, he laboured under great difficulty of breathing, and after death he was found to present a marked lateral curvation of the spine.

On examination, old adhesions and some fluid were found in the pleural cavities. The larynx and trachea were healthy. The lungs were congested and sparingly crepitant, and the branches of the pulmonary artery were atheromatous.

The abdominal organs were generally healthy, but the liver was large and firm, and, together with the stomach, intestines, and kidneys, congested.

The heart weighed twenty ounces and a quarter avoirdupois. There was considerable hypertrophy and dilatation of the ventricles, and especially of the right. The pulmonic valves were somewhat thickened, and the artery was rather large, and gave origin to the ductus arteriosus in the usual situation; the duct was pervious, so as to admit of the passage of a large writing quill, and opened into the descending portion of the aortic arch at a point three lines below the origin of the left subclavian artery. The aorta was throughout its upper portion unusually small, and after giving off the left subclavian artery, diminished still further in capacity, and again expanded beyond the point of entrance of the ductus arteriosus. At its commencement, the aorta had a circumference of thirty French lines (67·5 mm. 2·66 E. in.); beyond the origin of the left subclavian artery it was only capable of transmitting a ball measuring twenty-four French lines in circumference (54 mm. 2·13 E. in.), and below the entrance of the ductus arteriosus it again expanded to thirty French lines (67·5 mm. 2·66 E. in.).

In cases like that now described, the occlusion of the duct must have been prevented by the existence at the time of birth, of unusual narrowness of the aorta distal to the left subclavian artery, owing to the imperfect evolution of that portion of the vessel from the branchial arches. This form of malformation is therefore closely allied to the cases of constriction or obliteration of the aorta which are met with in the same situation in after-life, and it only differs from them in the period at which the obstruction occurs. In the cases in which the ductus arteriosus is found open, considerable narrowing must have existed before birth; where, on the contrary, the duct is closed, the constriction must have been then only slight, though it may subsequently have become very marked or have even led to the entire obliteration of the vessel.[1]

In some instances, also, the arterial duct and foramen ovale remain open without there being any cause to which the defects can be assigned. Thus, in a very interesting case related by Mr. Burns, a patient who had presented cardiac symptoms all his life, and had been slightly cyanotic and had suffered from attacks of dyspnoea, lived to about the age of forty; and after death the foramen and duct were both found open, without, apparently, any obstruction at the pulmonic orifice or in any other situation, to which the condition could be ascribed.[2]

In a case related by M. Huss, there was a direct communication between the left branch of the pulmonary artery and the aorta, and the aortic orifice was contracted and the valves cartilaginous. The patient was nineteen years of age, and the opening was supposed to have resulted from recent disease.[3]

---

[1] Path. Trans., vol. xiii. 1861–62, p. 38. Preparation in St. Thomas's Museum.
[2] Diseases of the Heart, p. 16.

[3] Gaz. Méd. de Paris, 1843, p. 91. A curious anomaly is described in the Path. Trans. by Dr. Wilks, in which a direct opening existed between the trunk of the pulmonary artery and the ascending aorta, vol. xi. 1859–60, p. 57.

2. DEFECTS WHICH DO NOT INTERFERE WITH THE FUNC-
TIONS OF THE HEART AT THE TIME OF BIRTH, BUT
MAY LAY THE FOUNDATIONS OF DISEASE IN AFTER-
LIFE.

IRREGULARITIES OF THE VALVES.

It has been previously mentioned that the valves at the
orifice of the pulmonary artery are frequently found defectively developed, in conjunction with other serious deviations
from the natural conformation of the heart; and the various
forms of irregularity which they present have been described.
The semilunar valves at the aortic orifice and the auriculo-
ventricular valves, have also been shown occasionally to present
morbid conditions under similar circumstances. Of these
forms of defect, when existing in cases in which the heart
is otherwise well formed, I proceed to speak more fully.

Malformations of the Semilunar Valves.

Deficiency in the number of the segments.—1st. When
the number of the semilunar valves is defective, there may
be only one curtain, which is stretched across the orifice
or protruded forwards in the course of the vessel, so as
to assume a funnel shape. When this is the case the fold
displays, on its upper side, three distinct septa or frena
(or, as they have been termed by John Hunter, *cross-
bars*[1]), dividing the same number of deeper or shallower
sinuses, and indicating the former lines of separation of the
several valves by the fusion of which it is probably formed.
The aperture is generally situated in the centre of the fold,
and has usually a triangular shape. This condition I have

---

[1] Descriptive Catalogue of Pathological Specimens contained in the Museum of
the Royal College of Surgeons, vol. iii. 1848, p. 199. Note in reference to
No. 1548, quoted from John Hunter's MS., "Dissections of Morbid Bodies." The
specimen here described consists of the semilunar valves at the commencement of
the aorta much ossified, and was removed from the body of a female forty
years of age. It is figured in Baillie's Illustrations of Morbid Anatomy, fasc. 1,
plate 2, fig. 3; and similar specimens are engraved in Carswell's Pathological
Anatomy, Hypertrophy, plate 2, fig. 4; Hope's Illustration of Morbid Anatomy,
fig. 74; and Cruveilhier's Anatomie Pathologique, liv. 28, plate 4, fig. 2.

already described as common in the valves of the pulmonary artery, and as often co-existing with other serious defects in the development of the heart; but it also, though more rarely, occurs in the aortic valves, and in hearts otherwise well formed.[1]

2ndly. In by far the most frequent malformation of the semilunar valves, there are two segments; and the deficiency is apparently due to the adhesion of the contiguous sides of two of the three original valves, and the atrophy of the corresponding angle of attachment. The former separation of the fused valves is generally indicated by the disproportionate size of the united segment; by the existence of a frenum or band dividing it more or less completely on the upper or arterial side; and, usually also, by a slight sulcus, running across its ventricular surface to the free edge and often terminating in a small notch. The attached margin of the united curtain has also generally the form of two more or less distinct crescents.[2] This kind of defect is frequently seen both in the aortic and pulmonic valves; and occurs, separately or in the two sets of valves, both in conjunction with obvious defects in the conformation of the heart, and in cases where the organ is well formed.

By most pathologists this condition of the valves has either been referred to the occurrence of disease in after-life; or to the breaking down of the angle of attachment from injury during violent muscular efforts. But, while both these causes may occasionally produce the blending of two curtains into one, the defect described is shown to be of congenital origin, at least in many instances, by its co-

---

[1] See plate 2, figs. 1 and 2; plate 3, figs. 1, 2, 4 and 5; plate 4, fig. 1.

[2] See plate 1, fig. 2; plate 2, fig. 4; plate 4, fig. 2; plate 6, fig. 2; and plate 8, fig. 1. The last drawing shows a case of fusion of two of the aortic valves, from a child which died when ten weeks old. There was also a contraction of the aorta distal to the origin of the left subclavian artery, and the descending aorta arose in great part from the pulmonary artery. See also fig. 2, plate 8, which shows the union of two of the aortic valves, from a boy aged 15, who was killed. Both these preparations are contained in the Victoria Park Museum, and are numbered B 17 and B 14. This defect in the valves is figured by J. F. Meckel, De cordis conditionibus abnormibus, Dissertatio Inauguralis, Halæ, 1802, tab. 2, figs. 2 and 3.

existence with other obvious malformations, and by its occurring in children which survive birth only for a short period, and even in fœtuses which have never breathed. It seems probable that the inequality in the size of the two existing valves, indicates the period at which the segments forming the united curtain become adherent. When the two segments are of nearly equal size and the frenum and sulcus very imperfectly marked, the fusion has probably occurred at an early stage of fœtal existence; when, on the other hand, there is considerable difference in size and the appearances of former separation are more distinct, the union may have taken place at a more recent period.

3rdly. The third and least frequent form of defect is that in which there are two large segments, with a small rudimentary valve interposed between them. This condition probably results from one of the segments becoming the seat of disease during fœtal life or early infancy, and so having its further development arrested.[1]

In most of the cases of deficiency in the number of the semilunar valves which I have examined, proofs have been presented of the originally triple form of the apparatus. It is not however to be denied, that cases occur in which no such evidence of former division can be detected; but the absence of such signs does not show that they have never existed. For if, as we occasionally see, two valves may be of nearly equal size and the indications of the former division of one of them may be only traceable on very careful examination; or if, when a rudimentary valve exists, it may be extremely small and imperfect, we can readily understand that, in some cases, all traces of the original condition of the valves may disappear, though the mechanism of the malformation may have been similar.

---

[1] See plate 8, fig. 3, from a specimen in St. Thomas's Hospital Museum, removed from a man sixty years of age, under the care of Dr. Bennett. The larger valves are extensively diseased. Notes of this case are published in Pathological Transactions, vol. iii. p. 289. A somewhat similar specimen is contained in Dr. Baillie's Museum, in the possession of the Royal College of Physicians, 4 A 13.

The forms of defect now described, though not necessarily interfering with the functions of the heart, do, in a considerable number of cases, lay the foundations of disease in after-life.

1st. When the whole of the segments are fused together or when two only are united, the valves not unfrequently become the seat of chronic inflammation, by which they are rendered thick and unyielding and often become extensively ossified; thus inducing, first obstruction to the flow of blood from the ventricle into the aorta, and then incompetency, by which the blood which has entered the aorta is allowed to regurgitate into the ventricle. This process is often so slow in its progress, that the ventricle accommodates itself to the additional exertion required; and the disease becomes a source of manifest evil only after the lapse of many years. This occurs when the valves have become so immovable as not to allow of the transmission of a sufficient column of blood into the aorta, or so much reduced in size as not to be capable of preventing its return; when the ventricle is no longer capable of performing the extra labour required to overcome the obstruction; or when the effects of more acute disease are superadded to the original valvular defect.

2ndly. When two of the segments are united, the curtain in the line of fusion or of former separation, being thickened and often indurated, is less readily extensible than the other parts of the united fold. Thus that portion often does not expand with the progress of growth so as to accommodate itself to the natural enlargement of the orifice; but is as it were held back, and leaves an open space between its edge and the other valve, by which the blood regurgitates from the aorta[1] into the ventricle.

3rdly. The tendency to regurgitation in the latter class of cases is increased by the want of the support to the edge

---

[1] This defect in the valves is described and figured by Dr. Brinton, as existing in a specimen exhibited at the Pathological Society in 1855.—Path. Trans., vol. v. p. 72.

of the larger curtain which should be afforded by the corresponding angle of attachment. The edge of the valve
thus becomes retroverted, and regurgitation is readily occasioned. In the Edinburgh Monthly Journal of Medical
Science, for 1853,[1] I have reported several cases of disease
of the aortic valves which were supposed to have originated
in one or other of these ways; and I have more fully illustrated my views in the Croonian lectures delivered before
the Royal College of Physicians, in 1865.

Excess in the number of the semilunar valves.—Specimens
displaying redundancy of the semilunar valves less frequently
occur than those in which the number of the curtains is
defective; and while the latter condition is probably most
common at the aortic orifice, it is the valves of the pulmonary artery which appear to be most frequently in excess.
Of forty-one cases of irregularity of the semilunar valves
which I examined and reported upon at the Pathological
Society in 1851, nine only exhibited excess in the number
of segments, and in eight of them the irregularity existed at
the pulmonic orifice.

The chief forms in which this description of irregularity
existed in the specimens were :—

1st. Four segments; three of nearly equal size, and a
smaller one interposed between two others, and generally
imperfectly separated from one of those to which it is
adjacent.[2]

2nd. Four segments of nearly equal size, or with two
larger than the other two, and in either case with two of
the segments imperfectly separated.

3rd. Four segments; one distinct, and three more or less
blended together.[3]

4thly. Three or four segments of nearly equal size, with
one or two smaller curtains interposed between others, and

---

[1] On Malformation of the Aortic Valves as a Cause of Disease.

[2] Plate 8, fig. 4, from a preparation marked B 13 in the Museum at the Victoria Park Hospital. The valves are the pulmonic, and the subject of the peculiarity was a female seventy-five years of age.

[3] In a specimen in the Victoria Park Hospital Museum. It was removed from
a man aged forty-five, who was crushed to death.

imperfectly separated from those to which they are adjacent.[1]

When the number of the semilunar valves is in excess, the redundancy would appear to be due to the more or less perfect subdivision of one or more of the segments. Thus, in some cases, the curtains of the redundant valves are more or less blended with those adjacent to them; and in others the angle of attachment of one of the segments is wanting, and the septum interposed between the two sacs is defective, being either perforated by larger or smaller apertures, or almost entirely wanting.

In the absence of any exact information as to the process of development of the semilunar valves, it is not easy to offer a satisfactory explanation of the cause upon which the excess in the number of segments depends. The peculiar arrangement of the apparatus in some fishes, may, however, afford an indication of the mode in which the valves are formed in man. In the preparation of the heart of *Cephalopterus Giorna* (Cuv.), or American Devil Fish, contained in the Museum of the Royal College of Surgeons,[2] three muscular columns are seen to extend the whole length of the bulbus arteriosus, from its commencement in the ventricle to its termination in the branchial artery. At a short distance from the ventricle small folds project from the sides of each muscular column, so as to form six imperfect valves, and of these there are several rows. The folds become more distinctly valvular as they advance towards the branchial artery, till at the termination of the bulbus arteriosus, they appear as three well-formed semilunar valves. Each muscular column terminates at the base of one of these valves, so that it would seem as if the latter were formed by the approximation and blending of two of the folds, the muscular column becoming at the

---

[1] This condition is illustrated in plate 8, fig. 5. This is a good example of five pulmonary valves, from a preparation marked B 12 in the Victoria Park Museum. The specimen was removed from a female four and a half years old. Excess in the number of the pulmonic semilunar valves is figured by Meckel in tab. 2, fig. 1.

[2] No. 910 of the Physiological Series.

same time abortive. A very similar arrangement is found
in the bulbus arteriosus of other cartilaginous fishes, as
the Grey Shark, *Galeus Communis* (Cuv.), and the Skate,
Raia Batis (Linn.). It is possible that the semilunar
valves in man may be developed in a mode analogous to
this ; each curtain consisting originally of two portions,
which are ultimately blended together and lose their central
attachment to the sides of the orifice. This suggestion is
in accordance with the fact that when more than the proper
number of segments exists, the excess is apparently due
either to one or more supernumerary curtains being attached
to other segments, or to the imperfect division of some of
the curtains. If this supposition be correct, the increased
number of valves, so far from depending on excessive development, results from arrest of growth :—the condition
being thus analogous to other forms of malformation which
apparently depend on excessive development, though they
really indicate the arrest of the process at a rudimentary
stage ; as has been before shown to be the case when a
septum exists between the sinus and infundibular portion
of the right ventricle, producing the appearance of three
distinct ventricular cavities.

It is, however, quite possible that the development of the
valves may take place in a mode the very reverse of that
here supposed. They may originate in a folding together of
the lining membrane with a portion of fibrous tissue, so as to
form rings encircling the orifices, of which the edges and
the upper or arterial surfaces become subsequently looped
up and adherent at intervals to the sides of the vessels,
so as to form separate semilunar segments. If this be
their mode of development, the theory that the deficient
number of curtains is due to the adhesion of the contiguous
sides of separately formed valves, must be abandoned, and
the condition be ascribed to arrested growth at early periods
of fœtal life ; while the increased number of segments must
be regarded as due to redundant development. The fact
that in some cases of very defective conformation of other
parts of the heart, the semilunar valves are found to be

represented only by a circular fold of the lining membrane, is in favour of the latter view ; but the former seems to afford a better explanation of the state of the valvular apparatus in the majority of cases.

It does not appear that excess in the number of the semilunar valves materially interferes with the efficient performance of their functions.   It is possible that the curtains may less readily apply themselves to the sides of the aorta and pulmonary artery during the ventricular systole, when there are four or five segments, than when the usual number exists.   If so, the condition may be the cause of some obstruction to the flow of blood from the ventricle; but this is very doubtful, for the results of subsequent disease are certainly much less frequently seen in such cases than when the number of the segments is defective.   There is no reason to suppose that the condition is at all productive of incompetency in the valves.

The semilunar valves are also occasionally found to present a condition to which the term *atrophy* has been applied ; the segments, chiefly at the free edges, and especially towards the angles of attachment, presenting small apertures or spaces, in which the fibrous tissue is wanting and the continuity of the curtain is only maintained by the two folds of endocardium.   Sometimes, also, the endocardium is deficient in places, especially at the edges of the valves, so that fibrous bands are found extending from the angles of attachment to different parts of the free edges.   This condition is common in valves which are otherwise irregular, especially when the number of the segments is in excess, and will be found figured in the plates.[1]   In these cases, therefore, it is doubtless a congenital defect ; but it is probable that in some instances it may occur in after-life from the stretching and distension of the curtains.   It does not appear ordinarily to interfere with the action of the valves, and is therefore an unimportant change.

---

[1] Plate 8, fig. 5.

Malformations of the Tricuspid and Mitral valves.

The three segments of the tricuspid valve are occasionally found united together, so as to form a kind of membranous septum, stretched across the right auriculo-ventricular aperture and perforated in the centre by a larger or smaller opening, generally of a triangular form. This condition closely resembles the first description of malformation of the semilunar valves; and though usually considered as resulting from disease in after-life, is probably in some instances of congenital origin. It occasionally exists in combination with other decided malformations, as in in cases reported by Lallemand and Louis,[1] and one of those previously related ;[2] and when this is not the case there are often other circumstances which indicate its congenital origin. Two cases of this kind have fallen under my notice. In one of these, related by Mr. E. Pye-Smith, in the Pathological Transactions,[3] the patient had been ailing all her life, and, though thirty-seven years of age, did not appear to be more than fifteen or sixteen, and had never presented any signs of puberty. She was feeble both in body and mind. She had never had rheumatism, or other affection likely to have been complicated with disease of the heart. In the other case, the patient, who was under my own care at the Aldersgate Street Dispensary, was thirty-two years of age, and had suffered from two attacks of rheumatic fever, the first thirteen, the second two years before she fell under my notice. Since the former seizure her symptoms had been aggravated ; but she was distinctly stated to have laboured under palpitation and dyspnœa prior to the first attack, and indeed from her earliest infancy. In both these instances the mitral valves were also adherent, and the orifice greatly contracted ; and in a case recorded by Mr. Burns, and one referred to by Laennec[4] and regarded by him as possibly

---

[1] Mémoires ou Recherches Anatomico-Pathologiques. Communications des Cavités Droites et Gauches, obs. 6 and 10. See also references at p. 59.
[2] P. 91.　　　　[3] Vol. iii. 1850–51, 51–52, p. 283.
[4] Diseases of Chest, Forbes's translation, 4th edition, 1834, p. 574.

resulting from intra-uterine disease, a similar combination existed. If therefore it be correct to regard the fusion of the curtains of the tricuspid valve as of congenital origin, it becomes probable that some of the cases of union of the folds of the mitral valve, also depend on changes taking place during fœtal life. The only author who, so far as I am aware, has previously expressed this opinion, is Mr. Burns,[1] and though his remarks on the subject have received the confirmation of Dr. Farre, they have hitherto attracted little attention. The two folds of the mitral valve are, however, often found remarkably blended together, so that they form a distinct sac protruded forwards into the left ventricle, and perforated in the most dependent part by a small slit—the well-known " *button-hole mitral.*" This condition often co-exists, as in the cases referred to, with similar disease of the tricuspid valve, or with some other marked deviation from the natural conformation of the heart. The fusion of the mitral valves to the more extreme degrees is also very generally found in young persons, and occasionally in those who have been ailing from birth and have never had any serious attack of illness. I am thus disposed to believe that Mr. Burns' views are, at least in some cases, correct. This inference is corroborated by the facts that the auriculo-ventricular apertures are occasionally obliterated during fœtal life; and that both the tricuspid and mitral valves have been found to present evidences of recent inflammation in newly born children, and even in the fœtus.[2] It is true that generally in cases of great contraction of the auriculo-ventricular apertures, the patients have at some periods of their lives laboured under rheumatism, and that their symptoms have either originated or become seriously augmented subsequent to such attacks. This is,

---

[1] Diseases of the Heart, 1809, pp. 28 and 30, quoted by Farre, On Malformations, p. 42.

[2] See case by Dr. Massmann of Berlin, quoted in Arch. Gén. de Méd., 5ᵐᵉ série, t. v., 1855, p. 80, in which the mitral and tricuspid valves presented recent disease in an infant which lived only twenty hours. The sounds of the fœtal heart before birth had been replaced by a loud bruit.

however, as has been several times remarked, often the case in instances of undoubted malformation; and is in accordance with the general pathological law, that, whenever a part has once been the seat of disease, it is ever after prone to take on similar action, during which any injuries which it may have sustained in the first attack become greatly aggravated. These remarks are, however, thrown out rather as suggestions to be confirmed or disproved by future observations, than as established views.

The mitral and tricuspid valves also occasionally present a condition similar to that which in the semilunar valves has been termed "*atrophy*." The curtains are found unusually short, and display portions in which the fibrous tissue is wanting, or the valve is perforated from the giving way of the two layers of endocardium. This cribriform condition is chiefly seen towards the free edges of the valves, and especially of the attached curtain of the mitral. It is probably sometimes a congenital defect, and in other cases occurs in after-life from the yielding of the folds when unduly stretched. It may possibly, in some cases, be a cause of incompetency.

The free fold of the mitral valve is occasionally found to have a rounded aperture in it, as if a portion had been punched out. These apertures vary from the size of a pea to a bean, and have in some instances been supposed to be congenital defects. They are, however, more probably the result of disease in after-life.[1]

DISPROPORTION IN THE CAPACITY OF THE CAVITIES, ORIFICES, AND VESSELS; AND DEFECTS IN THE SIZE AND FORM OF THE HEART.

Among malformations which may predispose to disease in after-life, may be classed the cases in which the walls and cavities of the ventricles are disproportionate to the size of

---

[1] A specimen of the kind was exhibited by Mr. Prescott Hewitt, at the Pathological Society, and is described in the Transactions, vol. iii., 1850-51, 1851-52, p. 78.

the orifices and vessels—conditions which have been supposed by Corvisart, Laennec,[1] and some others, to be congenital defects. There is, however, reason to doubt the correctness of this view; the disproportion referred to, if it exists, should rather be regarded as the result of irregular growth after birth.[2]

It is stated that the heart has in some cases been found too small, and the condition has been supposed to be congenital. Otto[3] informs us that he has met with very small hearts in cases in which there were other malformations, as well as in infants and grown-up persons in whom the organ was otherwise well formed; and he refers to similar observations recorded by Kerckring, Morgagni, Kreyzig, &c. Though, however, my attention has for many years been directed to the pathology of the heart, I have met with no case in which there was reason to believe that the organ had been congenitally defective in size. In after-life the heart is not unfrequently found very small. I have weighed hearts both of men and women which were only 5 or 6 oz. in weight (the average weight in adults being $9\frac{1}{2}$ and $8\frac{3}{4}$ oz. respectively), but these were all cases of emaciation, and the small size of the heart corresponded with the diminution in the general bulk of the body.

The external form of the heart has also been found in some cases irregular. Thus it has been mentioned that when the septum of the ventricles is defective, the aorta arises from the right side, and the pulmonary orifice is constricted or obliterated, the organ becomes much wider than usual, and has, indeed, the quadrangular form of the heart in the chelonian reptiles. When, on the other hand, the heart gives off only one vessel, and the septum of the ventricles is wholly or very considerably deficient, as in the

---

[1] Treatise on Diseases of the Chest, trans. by Forbes, 4th ed., 1834, p. 543.

[2] See Observations on Certain Diseases Originating in Early Growth, by Dr. Barlow, in Guy's Hospital Reports, vol. vi. 1841, p. 235, and vol. vii. 1842, p. 467. For further remarks on malformation as a cause of subsequent disease see the Croonian Lectures, 1865, by the author.

[3] Pathological Anatomy, South's translation, p. 264, sects. 1 and 2.

case of Mr. Wilson, the organ has a peculiarly elongated form. In some cases also the form of the heart is irregular, without the organ being otherwise defective. Thus Bartholinus[1] reports that he has found the apex bifid, and the heart of an infant displaying the same condition was exhibited by M. Parise at the Société Anatomique of Paris in 1837.[2] Something of the same kind is occasionally seen in cases in which with other defects the right ventricle is very greatly enlarged.

---

[1] Paget on Congenital Malformations, Edin. Med. and Surg. Jour., vol. xxxvi. 1831, p. 281.

[2] Bulletin, 1837, p. 100.

IV.

IRREGULARITIES OF THE PRIMARY VESSELS.

TRANSPOSITION OF THE AORTA AND PULMONARY ARTERY.

IN this malformation the points of origin of the primary vessels are transposed, the pulmonary artery arising posteriorly from the left, and the aorta anteriorly from the right ventricle. In the former part of this work several cases have been mentioned in which, with very defective development of other parts of the heart, the orifices of the primary vessels have been transposed. Sometimes, however, the same imperfection occurs in hearts which are otherwise not materially defective, and indeed in organs which are well formed. The first case of this kind with which I am acquainted is that described by Dr. Baillie in the second edition of the Morbid Anatomy, published in 1797,[1] and figured in the engravings, and of which the preparation is in the possession of the Royal College of Physicians.[2] It occurred in the practice of Dr. Wollaston, of St. Edmundsbury. The child died when two months old, and was remarkably livid and cold during life, but did not suffer from dyspnœa. The arteries were transposed, the ductus arteriosus was sufficiently large to admit a crow-quill, and the foramen ovale was a little more closed than in a new-born infant. The heart was of a natural size for a child of the age, and was otherwise healthy. In 1811 Mr. Lang-

---

[1] P. 38. Works by Wardrop, vol. ii. p. 36. Engravings, plate 6, fasciculus 1st.

[2] Numbered 4 A 7.

staff[1] related a similar case, in which the infant immediately after birth was of a dark purple colour in the face, and of a brownish black in other parts of the body. When three weeks old he began to suffer from attacks of dyspnœa and died suddenly in one of them seven weeks after. The condition of the heart was very similar to that in the case of Dr. Baillie. In 1814, Dr. Farre[2] recorded a case in which the child lived five months; and since that time others have been reported by Wistar,[3] Gamage,[4] Tiedemann,[5] Duges,[6] Coliny,[7] King,[8] Walshe,[9] &c.; and in 1851, a specimen of this kind of malformation was exhibited at the Pathological Society, by Dr. Ogier Ward,[10] which I had the opportunity of examining. It was removed from a child which was livid from the time of birth, had a feeble cry, took the breast with difficulty, and was sleepy and had irregular respiration. It died on the 18th day. The aorta arose

---

[1] London Medical Review, vol. iv. 1811, p. 88, quoted in Farre, on Malformations, p. 28.   [2] On Malformations, p. 29.

[3] System of Anatomy, 1814. Philadelphia, vol. ii. p. 78. In a child two and a half years old.

[4] New England Journal of Med. and Surg., 1815. Boston, vol. iv. p. 244. In a child fifteen weeks old.

[5] Zeitschrift für Physiologie, Heidelberg, 1824, Band 1, p. 111, with plate. The infant lived twelve days.

[6] Journal Gén. de Méd., t. ci. 1827, p. 88, in a child four or five days old.

[7] Arch. Gén. de Méd., 2me série, t. v. 1834, p. 284. In a male infant, which lived two years and seven months.

[8] Med. Gaz. 1841. Child lived two years and nine months.

[9] Med.-Chir. Trans., vol. xxv. 1842 (N. S., vol. vii.) p. 1, in a male infant which lived ten months.

See also Martin, Müller's Arch. für Anat. und Phys., 1839, p. 222, with plate. In a child which lived ten weeks.

Johnson, American Journal of Med. Sc., vol. xlvi. (N. S. vol. xx.) 1850, p. 370. In a male infant which lived two months.

Ducrest, Arch. Gén. de Méd., 3me et nouvelle série, t. ix. 1840, p. 76. In a female which lived ten hours.

Stedman, Lancet, 1841–42, vol. i. p. 645, with sketches. In a female child which lived seven months and eight days.

Stoltz, Gaz. Méd. de Paris, xxiime année, 1852, 3me série, p. 154, quoted from Gaz. Méd. de Strasbourg; two cases; first case, an infant which lived five days; second case, an infant which lived thirty-seven days.

[10] Path. Trans., vol. iii. 1851-52, 1852-53, p. 63.

from the right ventricle and the pulmonary from the left; the septum of the ventricles was entire, and the duct and foramen were both pervious.   The child died of congestion of the brain, and the lungs were found engorged and imperfectly expanded.   At about the period at which the first edition of this volume appeared, a case of transposition, with an able memoir on this form of defect, was published by Professor Mayer, of Zurich.[1]   In this memoir, the author includes various cases which have been previously alluded to in this work under other heads, together with four cases in addition to his own, which I had not previously noticed.   More recently, in 1863, an instance of the same anomaly was published in the Medico-Chirurgical Transactions, by Dr. Cockle.[2]

Transposition of the primary vessels may occur under different circumstances and in association with various other anomalies in the development of the heart.   The points of origin in the arteries may be reversed, either in combination with a corresponding misplacement of the heart and other viscera, as in the instance related by Mr. Gamage; or where the heart occupies its natural position, as in by far the larger number of cases on record.   It may be found in hearts which are otherwise very imperfectly formed, the septa of the ventricles or auricles or both being largely defective, as

---

[1] Arch. für Path. Anat. und Phys., von R. Virchow. Berlin, 1857, 12ter Band, p. 364.   The case of an infant four weeks old, cyanosed from birth.   The foramen ovale was open to the extent of two lines.   The ductus Botalli capable of admitting a thick bristle, the septum of the ventricles entire.   The cases quoted are those of Kiel (in an infant which survived thirty-five hours.   The foramen ovale was not entirely closed, and the ductus Botalli tolerably wide. The septum of the ventricles entire.   Wurtzburg, 1854).

Beck (seventy-five hours.   Foramen ovale open, ductus Botalli admitted a sound, septum of ventricles entire.   1846).

D'Alton in a cyanotic female two years old.   Foramen ovale open so as to admit the finger, septum of ventricles open, and ductus Botalli obliterated. Bonn, 1824. Figured by Förster, Taf. xviii. figs. 14 and 16.

Friedberg; foramen ovale open, ductus Botalli and septum of ventricles closed.

[2] Vol. xlvi. p. 193.   In a male child which survived two years and eight months, and was cyanotic.   Septum of ventricle entire, foramen ovale completely open, ductus arteriosus obliterated.

in the cases of MM. Martin[1] and Thore,[2] and Dr. Dickenson,[3] and in a preparation contained in the Museum of St. Thomas's Hospital;[4] when the septum of the ventricles is only defective to a slight extent, as in the cases of Coliny, Stedman, and King; or when the septum of the ventricles is entire, as in the cases of Baillie, Langstaff, and Farre; and Dr. Chevers states that in a preparation of the kind contained in the Museum of Guy's Hospital, the ventricles are completely divided. The last is, indeed, by far the most common condition.

In most instances the arteries only are misplaced; but in some cases, as those of Farre, Gamage, Walshe, and Stoltz, the ventricles also, as indicated by their relative size and by the form of the auriculo-ventricular valves, are transposed; and yet in others, the arteries and veins are irregular. Thus, in the case related by the late Mr. T. W. King, there were only two pulmonary veins; in that of Dr. Walshe, there were two carotid and two subclavian arteries; and in that of Mr. Gamage, the aorta followed an unusual course, making its turn to the right, passing down on the right side of the spine, and giving origin to the arteria innominata on the left side. The venæ cavæ were also inserted into the left auricle. In a case not included in the previous enumeration, reported by M. Boyer,[5] the viscera generally were reversed, the venæ cavæ entered the left auricle and the pulmonary veins the right. The aorta arose above an aperture in the septum cordis, so that it communicated with both ventricles; and the pulmonary artery was connected with the right ventricle. The foramen ovale was open, but the ductus arteriosus was not pervious through its entire course. The infant which was the subject of the malformation lived two months and had been cyanotic during life. A very similar case, before referred to, is related by Dr. Worthington.[6] The insertions of the

---

. [1] Referred to at p. 25.   [2] P. 25.   [3] P. 97.   [4] P. 25.
   [5] Arch. Gén. de Méd., 4ᵐᵉ série, t. xxiii. 1850, p. 90; and Gaz. Méd. de Paris, t. v. 1850, p. 292. This specimen was exhibited at the Acad. de Médecine.
   [6] American Journal of Medical Sciences, vol. xxii. 1838, p. 131; p. 96 supra.

cavæ and pulmonary veins were reversed, the pulmonary artery arose from the left ventricle, and the aorta was connected with both ventricles. There was no ductus arteriosus, but the foramen ovale was widely open, and the valves at the orifice of the pulmonary artery were wanting. The anomaly occurred in a female infant, twenty-two months old, which was cyanosed from birth.

In nearly all cases of this description of malformation, the foramen ovale is open to a greater or less extent, and generally also the ductus arteriosus is pervious; but in some instances the duct has been found occluded. This occurred in three out of four cases in which the septum of the ventricles was imperfect; and even when the ventricular cavities are completely separated, the duct is sometimes very small, and in four cases it is reported to have been entirely obliterated. In reference to one of these cases—that related by Mr. Gamage—Dr. Chevers suggests that there probably existed some communication between the pulmonary veins and the systemic auricle, or some other channel through which the vessels communicated; but in the case of Mr. Wistar it is expressly stated that there was no communication between the aorta and pulmonary artery by the ductus arteriosus, and that the veins entered the auricles as usual. In both instances the foramen ovale was largely open. In Dr. Cockle's and D'Alton's cases also, the only communication between the two sides of the heart was through the largely open foramen ovale. All the little patients were cyanotic, and they survived respectively fifteen weeks, two years and six months, one year and six months, and two years and eight months.

In 1854, I exhibited at the Pathological Society, a heart obligingly sent to me by Mr. Wordsworth, which had been removed from an infant under the care of Dr. Hess. The specimen displayed transposition of the aorta and pulmonary artery in conjunction with other defects.

Case XVII.—*Transposition of aorta and pulmonary artery ;
septum of ventricles largely open ; foramen ovale closed ;
ductus arteriosus probably open.*[1]

The following report of the case during life was furnished
by Dr. Hess :—

" The child was a strong, well-developed boy, but had a
cough, with hurried breathing and palpitation from birth.
He had repeated paroxysms of dyspnœa, some of which
continued for three hours ; and during the attacks he was
slightly livid in the face, but not more so than is common
in severe bronchitis.   He gradually grew weaker and
thinner, notwithstanding that the digestive functions were
unimpaired ; looked old in the face, and at last was a complete skeleton.   The extent of the dulness on percussion in
the region of the heart was much increased.   The heart's
sounds were clear, and there was no murmur.   The boy
died suddenly, when eight months old, in coma preceded
by convulsions, which came on after one of the asthmatic
paroxysms."

The heart was examined by myself.   It had a quadrangular
form.   The auricles were completely separated, the foramen
ovale being entirely closed by its valve.   Both auricles
opened into the left ventricle—the right auricle on the
right, the left auricle on the left side of the cavity.   The
right auriculo-ventricular aperture was of large size, the
left was much smaller.   The left ventricle was very large ;
and, at its upper and posterior part, gave origin to the
pulmonary artery.   The right ventricle was a small rudimentary cavity from which the aorta arose, and which
communicated with the left ventricle by a crescentic opening ten lines in circumference (22·5 mm. ·88 e. in.).   The
valves, at the orifice of the pulmonary artery were of very
unequal size.   The ductus arteriosus was not retained in
the preparation, but, from the size of the pulmonary artery,
the passage had probably been pervious.

---

[1] Path. Trans., vol. vi. 1854–55, p. 117.

The case is remarkable from the circumstance that the foramen ovale was entirely closed, but the peculiarity was, doubtless, due to the free communication which existed between the two ventricles.

The mode of formation of this case of malformation appears to have been a very complicated one. The transposition of the main arteries depended on the irregular evolution of the aorta and pulmonary artery from the primitive vessel and branchial arches; and the connexion of both auricles with the left ventricle may be ascribed to the right ventricle having been completely divided at the point of union between the sinus and infundibular portion ;—the latter giving origin to the aorta and opening from the left ventricle, while the sinus had become united with the left ventricle from which the pulmonary artery arose.

OTHER IRREGULARITIES OF THE PULMONARY ARTERY AND AORTA.

Allusion has already been made to the premature obliteration of the ductus arteriosus, as causing permanent smallness of the pulmonary artery. The vessel may also present other irregularities. Thus, it may be divided at a lower point than usual, as in a case observed by M. Cassan,[1] where the trunk was only three lines in length; in that of Mr. Bloxham,[2] in which it did not exceed one line; and a case of my own, in which it was about half an inch long.[3] In all these cases there was no trace of the ductus arteriosus.

The pulmonary artery may also arise by two distinct roots, as in a case recorded by Kerckring,[4] and before referred to, as an instance of duplicity of the right ventricle. In other instances the pulmonary artery, instead of being pro-

---

[1] Arch. Gén. de Méd., t. xiii. 5me année, 1827, p. 82.

[2] Med. Gaz., vol. xv. 1835, p. 435. In a female child which lived three years; the pulmonary artery was very small; the aorta arose from both ventricles, and the foramen ovale was imperfectly closed.

[3] Case xi. p. 100.   [4] Spicel. Anat., Amstelodami, 1670, f. 139, obs. 69.

longed through the ductus arteriosus into the descending
aorta, forms the left subclavian artery, as in the cases of
Calliot and Obet, Holst, Hildenbrand, &c.[1]   In yet other
cases, as in that of MM. Breschet and Martin, there may
be two communicating arteries in the place of the duct.[2]
In a case related by Mangas d'Angers, in addition to the
pulmonary supply being derived, as usual, there was a third
pulmonary vessel which arose from the abdominal aorta
near the cœliac axis, and passed through the œsophageal
aperture in the diaphragm, to be distributed by two branches
to the lungs.[3]   Cases have already been mentioned in which
there were supplementary or vicarious branches arising from
the aorta and distributed to the lungs.

In a specimen described by Hall and Vrolik, in which the
pulmonary artery was obliterated at its origin, there is said
to have been a second vessel which passed from the right
ventricle to the aorta.[4]

A case is mentioned by M. Bertin[5] in which the aorta
arose singly, but then divided into two branches and again
united to form the descending aorta.   The most frequent
and important irregularity is, however, that in which the
aorta, after having given off the vessels to the head and
upper extremities, becomes greatly contracted or obliterated
or is entirely wanting; so that the inferior parts of the
body derive their supply of blood from the pulmonary
artery through the ductus arteriosus.

Descending aorta given off from the pulmonary artery.

This form of irregularity, in which the descending aorta
is said to be given off from the pulmonary artery, was first

---

[1] Pp. 57 and 58 supra.   [2] P. 25.
[3] Journal de Méd. Chir. et Ph., t. iii. Paris, année 10, p. 453.   [4] P. 74.
[5] Maladies du Cœur, 1824, p. 433.   The anomaly was found by J. E. Bertin,
in a child ten or twelve years old.

A case has been more recently published by Bouillaud, in which the aorta is
said to have been double.   From the description given of the specimen, it may,
however, be doubted whether this was not a case of dissecting aneurism.—Arch.
Gén. de Méd., 1ᵐᵉ série, t. xv. 1847, p. 248.

noticed as existing in two preparations described by Dr. Farre, and contained in Sir Astley Cooper's museum. The preparations are now preserved in the Museum of St. Thomas's Hospital, and have been before quoted as examples of deficiency in the septum interposed between the left and the infundibular portion of the right ventricle, immediately below the origin of the pulmonary artery. They have on this account been described by Dr. Farre as instances of the pulmonary artery deriving its origin from both ventricles.[1]

The child from which one of these preparations was removed—that numbered in the Museum LL 65—is stated to have presented nothing unusual till a fortnight after birth, when it began to breathe quickly, and to waste; the heart pulsated strongly; the skin was always pale, and the hands and feet cold; the lower extremities, and occasionally the face, became œdematous, and the child died suddenly at the age of eight months. The ascending aorta divided into the arteria innominata, the left carotid, and the left subclavian arteries, and terminated in a very small trunk, which was continued into the descending aorta. The pulmonary artery was larger than the aorta, and arose from both ventricles (or more properly, arose from the right ventricle, but communicated with the left through the aperture in the septum), and, through the medium of the ductus arteriosus, gave off the descending aorta. The foramen ovale was much dilated. In the other case, the preparation of which is numbered LL 66,[2] the child during life had hurried and panting respiration, was of a dark purple colour, and died convulsed when nine days old. As in the other case, the aorta gave off the usual branches at the arch, and then formed a small trunk, which proceeded towards the descending aorta; but its cavity became impervious before it united with that

---

[2] The specimen which I have examined appears not to agree with the report given by Dr. Farre as to the vessel being impervious between the left subclavian artery and origin of the ductus arteriosus, though certainly it is of very small size.

vessel. The pulmonary artery arose from the right ventricle, above an aperture in the septum ventriculorum, and the ductus arteriosus was continued into the descending aorta. The valve of the foramen ovale was very imperfect, and the cells of the lungs were not fully inflated.

In 1834, M. Gibert exhibited at the Société Anatomique at Paris, the heart of a child which was born at the full period and lived twelve days, being during that time pale and cold, and suffering from difficulty of breathing. The aorta gave off the usual branches at the arch, and then diminished greatly in size, so that the descending aorta was derived chiefly from the pulmonary artery through the largely open ductus arteriosus. The foramen ovale was unclosed, and the septum of the ventricles incomplete.[1] A similar specimen was also exhibited before the same society in 1860, by M. Pamard. The child which was the subject of the malformation survived only thirty-six hours. It had also a club-foot and congenital hernia, but was not cyanotic. The portion of the aorta between the origin of the left subclavian artery and the ductus arteriosus was of very small size.[2]

CASE XVIII.—*Constriction of the aorta distal to the left sub-clavian artery ; ductus arteriosus largely open ; descending aorta chiefly supplied from the pulmonary artery.*

In 1847, I exhibited and described at the Pathological Society a preparation displaying this irregularity. The specimen was forwarded to me by the late Dr. G. A. Rees,[3] with the information that it was removed from an infant ten weeks old, which was thought to have been born at the eighth month. The child was feeble, and had much diffi-

---

[1] Bullet. de la Soc. Anat., année 14, 1839, p. 203 ; noticed also in the Proceedings, année 7, 1832, p. 108.
[2] Ibid., 2me série, t. v. p. 110.
[3] Path. Trans., vol. i. 1846–47, 1847–48, p. 203. The specimen is preserved in the Victoria Park Hospital Museum, and is numbered B 17. It is drawn in plate 8, fig. 1.

culty of breathing from birth.   It was never observed to be
livid in the face or extremities, and was indeed unusually
pale.   Large portions of the lungs were in the condition of
atelektasis.   The ascending portion of the aorta was unusually large; but after giving off the branches at the arch
the vessel diminished greatly in size, so that between the
origin of the left subclavian artery and the point of communication with the ductus arteriosus, it had scarcely half its
former calibre.   After the entrance of the duct, the aorta
again expanded, and it retained its increased size throughout
the thoracic portion.   The orifice and trunk of the pulmonary artery were large, and the duct was freely open and
led directly into the descending aorta.   The foramen ovale
was closed and the septum of the ventricles entire; and, except the aortic, the different valves healthy.   The aortic
was much smaller than the pulmonic orifice; and there appeared to be only two valves at its aperture.   The largest of
them was, however, obviously formed by the fusion together
of two of the segments.

An instance of malformation very similar to this is also
described in the Pathological Transactions by Dr. Chevers,[1]
and another was reported by Mr. Barrett[2] in the *Lancet* of
1835, which occurred in a still-born child.   The latter has
been before referred to as an instance of ectocardia abdominalis.[3]

In several instances, the defect at the commencement of the
descending aorta has been still greater, the ascending and descending aortæ having been found entirely disunited.   Thus,
in a child which survived birth only for a short time,
Steidele[4] found two vasa arteriosa to arise from the heart,

---

[1] Path. Trans., vol. i. p. 55. The infant lived eight hours, and died exhausted.   The septum of the ventricles appears to have been closed.

[2] Lancet, 1834 and 1835, vol. i. p. 349, in a fœtus which had also a large
congenital umbilical hernia with ectopia of the heart.   It is stated that the
ductus arteriosus was in connexion with the left subclavian artery. State of
for. ov. and vent. sept. not named. P. 7 supra.

[3] See also Förster, Taf. xix. fig. 4, from a preparation at Würtzburg.

[4] Hein, obs. 66. The other particulars of this case are not reported.

one of which—the pulmonary artery—furnished two branches to the lungs, and was then continued into the descending aorta ; while the other—the aorta—formed the carotid and axillary arteries.   A similar condition was also found in a fœtus of the ninth month, described by Mr. Struthers and Dr. Greig.[1]   In this instance, there was no connexion between the aorta and pulmonary artery, and the descending aorta was entirely derived from the latter vessel.   The aorta arose from both ventricles and the foramen ovale was slightly open.

In the Pathological Transactions for 1864, Dr. Wale Hicks described a specimen which occurred in a child which survived birth only for thirteen hours, and was very livid during all that time.   The aorta arose as usual and gave off the vessels to the head and upper extremities, while the descending aorta was derived from the pulmonary artery ; and there was no connexion between the two vessels.   The septum of the ventricles was entire, and the foramen ovale, though covered by its fold, was still pervious.   This case, therefore, differs from that of Struthers and Greig, in having had the septum of the ventricles complete, but probably agrees in this respect with that of Steidele.[2]

Cases of this description, in which the portion of the aorta between the origin of the left subclavian artery and the insertion of the ductus arteriosus is imperfectly developed, are closely allied to those in which, in persons dying in after-life, the same part is obstructed or obliterated.   They differ from them, indeed, probably only in the degree of constriction which exists at the time of birth. If that be great, the ductus arteriosus will remain open ; if only slight, the closure of the fœtal passage may not be prevented, but the defective state of the vessel may be the source of serious obstruction to the circulation in after-life.

---

[1] Monthly Journal of Med. Sc., vol. xv. 1852 (N. S., vol. vi.), p. 29, with woodcut.   See also Anatomical and Physiological Observations, by John Struthers, part 1, Ed. 1854, p. 75.

[2] Path. Trans., vol. xv. p. 85.   This specimen is figured in plate 7, fig. 3, of this work.

On this, as it has been termed *quasi malformation,* a memoir by the author was published in the British and Foreign Medico-Chirurgical Review in 1860,[1] in which forty instances of the kind were collected and analysed, of which ten were examples of entire obstruction, and thirty of more or less constriction. At the present time there cannot be less than forty-six cases on record, of which eleven are of the former and thirty-five of the latter class.

In these cases, though as before stated, the constriction at the time of birth may have been only slight, or, the coats of the vessel may only have been rigid and unyielding, so as not adequately to dilate with the progress of growth, the obstruction may be sufficient to interfere with the free transmission of the blood to the lower parts of the body. The heart, therefore, acts with increased power to overcome this obstruction, the left ventricle becomes enlarged and its walls increased in thickness; the ascending portion of the aorta dilates, and from the pressure of the blood, the coats in the contracted part become thickened and indurated; not unfrequently fibrinous material is deposited in the canal; and ultimately, as the result of changes occupying a long period, the calibre of the vessel may be entirely obliterated. A very remarkable example of this condition is preserved in the Museum of St. Thomas's Hospital, and has been described by Mr. Sydney Jones in the Pathological Transactions.[2] It was obtained from the body of a man forty-five years of age, sent to the hospital for dissection, and of whose previous state no account could be obtained. A case of nearly complete obliteration also occurred at St. Thomas's, in 1859, in the practice of Dr. Barker, and is described in the Medico-Chirurgical[3] Transactions. It occurred in a man twenty-four years of age, who was stated to have been in his usual health till eighteen months before his death. He died suddenly from the formation of a dis-

---

[1] Vol. xxv. p. 467.
[2] Vol. viii. 1856–57, p. 159. The body had been injected, and the collateral circulation is well shown in the preparation.
[3] Vol. xliii. p. 131.

secting aneurism, and the escape of blood from the sac into the pericardium.

DEVIATIONS FROM THE NATURAL ARRANGEMENT OF THE VENOUS TRUNKS.

The most frequent of these irregularities are the existence of two ascending or two descending cavæ;[1] of two or three instead of four pulmonary veins;[2] excess in the number of the veins;[3] or their irregular insertion, so that they may enter the vena cava[4] or be directly inserted into the ventricles. In some cases of transposition of the aorta and pulmonary artery, it has been mentioned that the veins also are reversed as in those of Boyer,[5] and Worthington,[6] in which the pulmonary veins entered the right, and the systemic veins the left auricle. The two sets of veins may also communicate with the same cavity. An interesting instance of irregularity in the course of the veins, occurred in the case of M. Raoul Chassinat, which has been previously quoted. There were only two pulmonary veins, and the vessel which conveyed the blood from the left lung entered the left auricle; while that for the right lung passed through the diaphragm and entered the vena cava ascendens above the point of insertion of the ductus venosus.[7] These irregularities are not, however, of sufficient importance to require further notice.

IRREGULARITIES IN THE CORONARY ARTERIES AND VEINS.

It has been already mentioned, that when the primitive vessel arising from the heart does not divide naturally into the aorta and pulmonary artery, the origins of the coronary arteries are irregular; those vessels being derived from the

---

[1] Kerckring, Spicel. Anat., obs. 29, f. 68.   Also specimen in St. Thomas's Museum, LL 67; Breschet, p. 23 and 25 supra.

[2] Foster, p. 16; Vernon, p. 17; and Mauran, p. 18.

[3] Sandifort, Obs. Anat. Path., lib. iii. f. 18 and 41; and lib. iv. f. 97.

[4] Mr. Wilson's case, p. 15 supra.   [5] P. 146 supra.

[6] P. 146 supra.   [7] P. 78 supra.

aorta some distance above its origin or from one of the primary branches. Thus, in the cases of Dr. Farre and Mr. Foster[1] they appear to have arisen from a single branch which took its origin from the concavity of the aortic arch. In the case of Mr. Clark and Professor Owen,[2] and in that of Dr. Vernon,[3] they arose by a common trunk from the innominate artery. In the report of Mr. Wilson's case, the point of origin of the coronary arteries is not mentioned, but in the preparation at the Royal College of Physicians, they appear to arise by a trunk, derived from the primitive vessel before its division into the aorta and pulmonary artery. In the case of Mr. Power, the coronary artery also apparently arises from the aorta or one of its branches.

In some cases in which the heart and vessels are otherwise well formed, the coronary arteries deviate from their natural position. Thus they may be excessive or defective in number. I have seen the anterior branch of the right coronary artery arise as an independent trunk from the right sinus of Valsalva; and in other cases, there is only one coronary artery which gives off the usual vessels as branches. This condition has also been noticed by Otto. In some cases where there are two vessels, the points of origin are irregular, both being connected with the same sinus; or they may arise from higher points than natural.

The coronary veins may also be irregular, both in number and connexion. Thus the vein may enter the subclavian vein or be inserted into the left auricle.

---

[1] Pp. 15 and 16 supra.   [2] P. 16.   [3] P. 17.

V.

MODE OF FORMATION, SYMPTOMS AND EFFECTS, MEDICAL MANAGEMENT, ETC.

1. MODE OF FORMATION.

IN all cases of malformation there must exist some primary deviation from the natural process of development upon which other secondary changes are dependent; and the different forms of anomaly may be divided into three classes, according to the nature of such primary defect. Those of the first class are all more or less closely allied, and depend on the process of development being arrested at different periods of fœtal life; so that the organ retains one or other of the forms which are proper to it in different stages of evolution.

In some cases, and especially when the heart consists of only two cavities, as in the instance described by Mr. Wilson, we cannot detect any cause to which the arrest of development can be ascribed; but in others, in which the growth of the organ has proceeded to a more advanced stage, we are frequently able to trace the circumstances which have prevented its further development. This will, however, be more apparent, if we reverse the course adopted in describing the various kinds of malformation, and trace the condition of the heart from the more perfect to the more rudimentary forms.

If, during fœtal life, after the septum of the ventricles has been completely formed, the pulmonic orifice should become the seat of disease, rendering it incapable of transmitting the increased current of blood required to circulate through the lungs after birth, the foramen ovale may, as was clearly shown by Morgagni,[1] be prevented closing; and if, as was further pointed out by Dr. Hunter,[2] the obstruction take place at an earlier period, when the septum cordis is incomplete, a communication may be maintained between the two ventricles. The same cause may also determine the permanent patency of the ductus arteriosus; for if during fœtal life the pulmonary artery be much contracted or wholly obliterated, the blood must be transmitted to the lungs through the aorta; and, unless the ductus arteriosus be itself obstructed, that vessel will necessarily become the channel by which it is conveyed. Similar effects would result from obstruction in the course of the pulmonary artery or in the lungs; in the right ventricle or at the right auriculo-ventricular aperture. So also, obstruction at the left side of the heart, as at the left auriculo-ventricular aperture or at the orifice or upper part of the aorta, would cause the current of blood to flow from the left auricle or ventricle into the right cavities, and thence, through the pulmonary artery and ductus arteriosus, into the aorta, and would equally determine the persistence of the foramen and duct or of an opening in the ventricular septum. The pulmonary artery and aorta would indeed appear to be either capable of maintaining for a time both the pulmonic and systemic circulations; and the necessary effect of the one vessel having the twofold function to perform, would be to give rise to hypertrophy and dilatation of the cavities of the heart more directly connected with it, and to the atrophy and contraction of those which are thrown out of the course of the circulation.

These effects of obstruction at the different apertures must vary according to the period of fœtal life at which the impediment occurs. If, as has been before shown, the pulmonary artery be obstructed before the complete

---

[1] Letter 17, arts. 12 and 13.
[2] Med. Obs. and Enq., vol. vi. p. 305.

division of the ventricles, the aorta may be connected with the right ventricle, and both the systemic and pulmonic circulation may be chiefly maintained by that cavity. If, on the other hand, the obstruction take place after the completion of the septum, the double circulation will be carried on by the left ventricle :—in the former case the left ventricle; in the latter the right, becoming atrophied. The degree of obstruction may also influence the course of the circulation and so affect the development of the heart. A slight impediment at or near the pulmonic orifice, while the growth of the septum cordis is in progress, will probably give rise to hypertrophy and dilatation of the right ventricle and to the persistence of a small inter-ventricular communication. More aggravated obstruction, on the contrary, may arrest the process of development and throw the maintenance of the circulation on the left ventricle.

The influence of obstruction at or near the pulmonic orifice or in some other portion of the heart, in modifying or arresting the development of the organ, is thus far capable of demonstration; but it is probable that similar causes may equally give rise to the more extreme degrees of malformation, in which one or other cavity retains its primitive undivided condition. For if obstruction taking place during the growth of the septum be capable of preventing its complete development; it may be inferred that impediments occurring at a still earlier period, may entirely arrest the formation of the septa, so as to cause the ventricle or auricle or both to remain single, or to present only very rudimentary partitions. It cannot, indeed, be disputed that in some cases, more particularly when the arrest of development is extreme, no source of obstruction exists to which the defect can be assigned; but it must be borne in mind that the absence of any obvious impediment to the circulation after the lapse of a considerable period, as in persons dying several years after birth, does not afford any proof that some obstruction may not have existed when the deviation from the natural conformation first commenced. On the contrary, as remarked by Dr. Chevers, the condition which at first sight appears least in accordance with the theory of obstruction,—that in which the pulmonary orifice and artery are dilated,—really affords evidence that some serious impediment must have existed in the lungs or elsewhere, though it may have entirely disappeared.

These inferences seem, indeed, so natural that I should not have thought it necessary to dwell upon them, especially as they have received the support of some of the ablest writers on the subject—Louis, Williams, Craigie, King, and especially Dr. Chevers—had it not been that different views have been advanced by M. Bouillaud, and more recently by M. Forget. The former author not only contends that the apertures in the septa are frequently due to disease in after-life; but ascribes the thickening and induration of the valves which most generally cause the obstruction at the pulmonic orifice, to inflammation occurring after birth, and induced by the entrance of aërated blood into the right ventricle. To these views there are insuperable objections. In a large proportion of cases, the pulmonary orifice or artery is so contracted as only to allow a small amount of blood to be subjected to the influence of the air, and thus the circulating fluid must be chiefly venous, and can have little effect even if it enter the right cavities. Very generally, also, the right ventricle is more powerful than the left, and the blood must flow from the right to the left side, instead of in the opposite direction as supposed by M. Bouillaud. The cause, therefore, which he regarded as producing the valvular defect cannot under ordinary circumstances operate. In many instances, also, there is conclusive proof, that the disease of the valves is congenital; for it has been found in infants which have died very shortly after birth, and even in fœtuses which have never breathed.[1] In other cases its

---

[1] I have myself found the valves adherent in a monstrous fœtus, still-born, and in an infant, a girl ten weeks old, in which the disease was evidently of old date. Specimens have also been obligingly forwarded to me by Mr. Obré, Dr. R. Quain, and Dr. Ingram, showing the disease in infants which died at the ages of six weeks and six months, and in which there was no appearance of any recent affection of the heart, and the children died suddenly without having been previously indisposed.—Path. Trans., vol. iv. pp. 96 and 101.

M

intra-uterine origin may fairly be inferred from the precise
similarity of the changes to those which are clearly congenital.

While, however, I differ in opinion with M. Bouillaud, as
to the relation which exists between the pulmonic disease
and the defects in the ventricular and auricular septa, and
as to the period at which the changes at the pulmonic orifice
occur, I fully concur with him in regarding the thickening
and induration of the valves, which is so frequently the
cause of obstruction, as due to inflammation.   Analogy
would lead us to expect that the fœtus in utero is liable to
diseases precisely similar in their nature and results to those
which affect the child after birth; and the correctness of
this inference is confirmed by clinical experience, for there
can be no doubt that both the peri- and endo-cardium are
occasionally the seat of inflammation during fœtal life.[1]

This view, however, is not without its difficulties.   In
after-life the valvular affection is most usually situated on the
left side of the heart, and when the valves on the right side
are also involved, they are usually much less diseased.
Whereas in cases of malformation it is on the right side
that the changes are chiefly found.   If, then, the process
at the two periods be identical, to what is the difference
in its seat to be assigned?   It cannot depend on the
blood which circulates through the right ventricle and
pulmonary artery being more stimulating than that which
enters the left ventricle and aorta.   So far as there is
any difference in this respect, as a large portion of the
blood which is conveyed to the right ventricle by the ductus
venosus and inferior cava, is transmitted through the foramen ovale to the left auricle and ventricle, did this cause
operate in the production of disease, its effects would be
manifested on the left side.   Neither is the function of the

---

[1] A case by Dr. Massmann, of Berlin, in which there was disease of the mitral
and tricuspid valves during fœtal life, has already been referred to. Billard also
relates that in an infant two days old he found strong adhesions between the
laminæ of the pericardium, evidently of old date, and regarded them as the
result of pericarditis during fœtal life.

right ventricle during fœtal life more active than that of the left; for the blood which enters the right auricle is equally distributed between the two ventricles. In the absence of any more satisfactory explanation, I am disposed to think that the more immediate connexion of the right ventricle with the circulation in the descending aorta and umbilical arteries, may explain the greater liability to disease at the orifice of the pulmonary artery ; for the circulation in the cord and placenta would appear to be more liable to temporary obstruction than that in the body of the fœtus itself. We know that in after-life the variable pressure of the blood in the arterial system is a fruitful cause of disease in the aortic valves.

The second form of malformation—that in which the vessels are misplaced or unusually distributed, must be ascribed to perversion of the process of development causing the improper division of the primitive vessel and the irregular evolution of the branchial arches.

The transposition of the aorta and pulmonary artery cannot be referred to any deviation from the natural position of the septum cordis. It is not simply that the pulmonary artery arises from the left ventricle and the aorta from the right; but the positions of the two vessels are reversed, so that the pulmonary artery originates from the posterior and upper part of the left ventricle, and the aorta from the anterior and upper part of the right ;—the lower portion of the latter vessel crossing the origin of the pulmonary artery in the same way that the pulmonary artery ordinarily embraces the origin of the aorta. The production of this form of malformation is to be ascribed to the irregular division of the arterial trunk, so that the branchial arches ordinarily associated with the portion of the vessel which forms the pulmonary artery, become connected with the aorta; while the arches which should be associated with the portion which becomes the aorta, are in connexion with the pulmonary artery. This deviation from the natural development may take place either after the septum of the ventricle is completed, or while the growth of the septum is in progress, so that the transposition may involve the ventricles as well as the vessels. The crossing of the aorta over the right branchus instead of over the left, must be ascribed to persistence of the portion of the branchial arch which forms the right aorta, while that which should form the left aorta has become abortive. So also the contraction or obliteration of the aorta beyond the left subclavian artery, which sometimes occasions the open state of the ductus arteriosus, is due to that part of the branchial arch having become abortive; and the existence of two arterial ducts or of two aortæ, must be referred to the persistence of portions of the arches which should have disappeared. Irregularities in the number or position of the large venous trunks admit of explanation in the same way.[1]

In the third class of anomalies, the irregularity in the conformation of the heart is due to the combination of both the previous forms of defect. Not only has the process of development been arrested at different periods of fœtal life, but the evolution of the organ itself and of its primary vessels has been otherwise irregular. Such is the nature of those cases in which, with great imperfection of the septum of the ventricles or auricles or both, the points of origin of the aortic and pulmonary artery are transposed, as in the instances quoted from MM. Breschet and Martin, Dr. Kussmaul and Dr. Dickenson, and the specimen described by myself which exists in St. Thomas's Museum.

In the descriptive portion of this work I have made no allusion to excessive development of the heart; for various malformations which have been regarded as examples of that condition, such as those in which the right ventricle is divided into two cavities, or where there are two aortæ, or a ductus arteriosus on each side, or two descending or as-

---

[1] This subject has been very ably illustrated by Dr. Turner, in a paper in the British and Foreign Medico-Chirurgical Review, to which I have pleasure in referring as exhibiting the latest views as to the development of the heart and large vessels, and so throwing much light on the mode of production of various forms of irregularity.—Vol. xxx. July to Oct. 1862, pp. 173 and 461.

cending cavæ, really result from defective development;
while others are met with only in monstrous fœtuses incapable of extra-uterine life, and are therefore of no practical importance, and as such would be beside my purpose.

The various facts which have been recorded show that in
the heart, deviations from the natural process of development may occur during all stages of fœtal life—before the
division of the cavities has commenced, after it has to some
extent progressed, and when it is entirely completed. It is
not, however, clear at what period the irregular development
most commonly begins. If inferences were to be drawn
from the published cases of malformation, it would appear
to take place most frequently during the later periods of
fœtal life. The correctness of such a conclusion may,
however, be doubted. The earlier the period at which the
process of development is deranged, the greater will probably be the defect, and the less readily will the system
accommodate itself to the change after birth. So that the
cases in which, with any marked deviation from the natural
conformation, extra-uterine life is maintained for a longer or
shorter period, probably constitute only a small proportion
of those in which the development is irregular. It seems
reasonable to suppose that during the earlier periods of fœtal
life, when growth is most rapid, the process would be most
liable to derangement.

In the production of defective development, sex has been
supposed to exercise an influence, and malformations are
certainly most common in males, though why it should be
so seems incapable of explanation. I find, however, that of
110 cases of malformation which I have collected, and in
which the sex of the subjects is recorded, 61 were males,
and 49 females, or 55·4 and 44·5 per cent. respectively.

The occurrence of accidents and strong impressions upon
the mind of the mother, are also supposed to conduce to
the irregular development of the offspring, and in many
cases such causes appear to have operated. In several
instances which have fallen under my notice, the mothers
of children labouring under malformations of the heart,

have assigned the defects in their offspring to strong mental impressions or shocks which they sustained during pregnancy; and there seems reason to believe that such causes, by deranging the maternal and indirectly the fœtal circulation, might produce the effects. In other instances, also, there has apparently been an hereditary predisposition to defective development of the heart, more than one child of the same parents having been affected.

SYMPTOMS.

The symptoms by which malformations of the heart are characterized, are referable to derangement of the circulatory and respiratory functions, and to the secondary disorders of the various viscera so induced. A child, labouring under serious malformation of the heart, generally presents at birth a very livid colour. The respiration is imperfect and difficult, and the heart is observed to beat violently. There may, however, be no unnatural appearance whatsoever at the period of birth, and the symptoms may only present themselves at the expiration of some months or even years. When once the signs of obstructed circulation have been manifested, they may either be permanent, or they may soon subside, and the child may acquire a healthy appearance, and thrive naturally. In either case, however, paroxysms generally occur, in which the breathing becomes extremely difficult, rapid, or gasping; the surface of the body, more especially of the face and extremities, acquires a dark, in some cases almost a black hue; and the infant indicates by its cries great distress. Not unfrequently it becomes convulsed, and when exhausted by its struggles, the symptoms subside, and disappear more or less completely. The paroxysms recur at longer or shorter intervals, and with varying degrees of severity, according to the extent to which the deviation from the natural conformation of the heart predisposes to embarrassment of the circulation. I find, by a careful analysis of 101 cases of malformation, characterized by well-marked symptoms during life, and in which the period at which the symptoms were first observed is reported, that in 74 cases they were first noticed either at the period of birth or very shortly after; while, of the remaining 27, in 15 they appeared before the expiration of the first year; in 1 at the sixteenth month; in 3 at two years; in 2 at three years; in 1 at three and a half years; in 2 at five years; in 1 at eight years; in 1 at thirteen years; and in 1 at fourteen years of age; and the period which elapsed might be still further extended. In cases, however, in which the symptoms are stated to have been long deferred, it may be doubted whether, during earlier life, there were really no indications of any irregularity about the heart. It is more probable that the period at which the signs are said to have been first manifested, was rather that in which they became so much aggravated as to attract attention. In two cases the symptoms supervened or became much more marked, after falls at five and fourteen years of age.

In some cases, in the intervals between the paroxysms, the action of the heart may not be materially deranged, the pulse being natural or only displaying some want of power. In others it may be constantly rapid and jerking; and in the paroxysms it frequently becomes intermittent, very irregular in force or frequency, or barely perceptible. The vessels in the neck may beat visibly, and there may be a marked venous pulsation. The impulse of the heart is usually powerful, and an unnatural murmur is generally heard. The condition of the respiration is liable to similar variations. In some cases there may be but little dyspnœa; in others the respiration is at all times rapid and laborious; and it is usually rendered much more difficult by any exertion or excitement. The lividity of the surface may be only slight, so that the lips may be a little purple and the nails discoloured; or even these symptoms may not be present, the child retaining either its natural appearance or being unusually pallid. In other cases, the surface generally is extremely livid, and the lips, hands, and feet of a deep purple colour: and the congestion of the capillary circulation is shown by the slowness with which the colour returns into the integuments, after they have been blanched by compression.   The peculiar tint of the surface varies from a slight rose or livid colour, to deep purple, blue, or black; the latter being most generally observed in the lips and beneath the nails.   The capillary vessels in the cheeks are distinctly marked, and not unfrequently those of the con-junctivæ convey dark blood.   The duration of the paroxysms varies from a few minutes to several hours, and they may recur very frequently, or only every two or three days, or still more rarely.   Sometimes the attacks are only brought on by active exertion, as when the child is washed or dressed, or on exposure to cold; but, in other cases, the slightest exertion, even that of taking the breast, or any trivial transition of temperature, will be followed by increased difficulty of breathing, palpitation, and lividity.   The nurse frequently finds out by experience some mode by which she can relieve the paroxysms when they occur, as by patting the back of the child, or laying it across the lap with its face downwards so as to compress the chest.   The dis-position of the infant is generally very irritable, and if it be thwarted the paroxysms are immediately brought on. Though occasionally the integuments become infiltrated, so as to give the child a gross appearance, more frequently it is much emaciated.   The abdomen is generally tumid and the head large.   In a case, however, which has been for some time under my care, in which I have reason to believe there is some serious deviation from the natural conforma-tion of the heart, the head is remarkably small.[1]

If the child survive the period of infancy the symp-toms continue similar in character.   From the defec-tive respiration the power of generating heat is very feeble, the extremities are cold, and there is peculiar susceptibility to changes of temperature, so that the child is liable, on the slightest exposure, to suffer from bronchitic attacks.   The

---

[1] This child has since died, and presented a remarkable example of atrophy of the brain, but the heart was well formed. See Path. Trans., vol. x. 1858–59, p. 15.

paroxysms of dyspnœa terminating in syncope and convulsions, are brought on by any unusual excitement of mind, by active exertion, by cold, or by any disorder of the stomach from indigestible food.   The children often learn to avoid these causes, so that they do not engage in play with others and prefer a warm fire-side.   In some instances they can arrest the progress of the attacks by lying down, or by compressing the chest in any other way against a resistent body.   Generally their mental power is very feeble ; and not unfrequently they suffer from pains in the head and occasional convulsive attacks ; and they are liable to hæmorrhages, and to unhealthy sores in different parts of the body, more particularly on the fingers and toes and around the anus.   Most usually they are thin, and the fingers and toes have a bulbous shape, and the nails are incurvated.   In some cases of malformation the emaciation is extreme.   Dr. Hunter[1] remarks, in reference to the boy whose case he has related, that " though he was remarkably thin, he had not the look of being emaciated by consumption ; on the contrary, it appears to be his natural habit. If a man had never seen any of the canine species but the bull-dog, for example, he would be struck at the first sight of the delicate Italian greyhound.   This young gentleman put me in mind of that animal, and when I looked upon his legs particularly, I could not but think of the legs of a wading water-fowl."   In females, if they survive to the period of puberty, the catamenial function is rarely established.

Dr. Farre made several observations on the temperature of persons in whom there were the signs of malformation of the heart, which seemed to show that, notwithstanding the sense of chilliness and susceptibility to cold, the body ordinarily possessed the average temperature.   I have also directed my attention to this subject, and have repeatedly noted the temperature under the tongue and in the axilla in cyanotic children, and especially in the boy who was the subject of Case II., without finding that the temperature

---

[1] Med. Obs. and Enq., vol. vi. 1784, p. 300.

differed materially from that of another child, quite healthy but somewhat younger, examined at the same time.

Cyanosis. Cyanopathia, Morbus Cæruleus.—There are few subjects in the range of medical science which have occasioned more discussion than the inquiry as to the immediate cause of the discoloration of the surface, which forms so marked a feature in most instances of malformation of the heart.

Morgagni, in describing the case to which I have several times referred, ascribed the marked cyanosis which had been observed during life to general congestion of the venous system, caused by the obstruction at the origin of the pulmonary artery, and this view was also adopted by Dr. Pulteney. Dr. Hunter, on the contrary, seeing that in the case which he has related the septum cordis was imperfect, and the aorta was supplied from both ventricles, so that a large proportion of the blood circulating in the body must have been venous, supposed that the livid colour the boy had presented during life, was owing to the intermixture of the currents of blood. These views have each since met with numerous supporters. The theory which ascribes the production of cyanosis to congestion of the venous system has been advocated by Louis, Ferrus, Cruveilhier, and Valleix in France; by Hasse and Rokitansky in Germany; by Joy in this country, and very ably by Stillé in America. On the other hand, the view which refers the discoloration to the intermixture of the venous with the arterial blood, has been supported by Gintrac and, with some modification, by Bouillaud and Forget, in France; by Meckel in Germany; by Lombard; and by Farre, Paget, Williams, Hope, Crampton, and Walshe, in this country. Corvisart and Laennec appear disposed to adopt the former explanation; and Dr. Chevers, while regarding the cyanosis as chiefly due to congestion, contends for the influence of the venous blood in the arteries, as increasing the intensity of the discolorization.

Gintrac, from an analysis of 53 cases, in all of which there was more or less intermixture of the currents of blood, inferred that cyanosis is dependent on the presence of venous blood in the general circulation, though he admitted that the intermixture was not always productive of cyanosis. Louis and Ferrus have dwelt more fully on the absence of any constant connexion between the intermixture of the currents and the existence and intensity of cyanosis. Dr. Stillé,[1] after a careful examination of a very extended series of cases of different forms of malformation of the heart, has shown, 1st, that cyanosis may exist without the intermixture of the currents of blood; 2ndly, that there is no just proportion between the intensity of the cyanosis and the amount of venous blood which enters the systemic vessels; 3rdly, that complete intermixture may take place without cyanosis being produced; and 4thly, that the variations in the extent, depth, and duration of the discoloration, are inexplicable by the doctrine of the intermixture of the currents. Of 77 cases which he collected and carefully analysed, he found the condition of the pulmonary artery reported in 62; and in 53 of these it was contracted, obstructed, or impervious; while in the remaining 9 cases, there were other conditions present which would give rise to congestion of the venous system. He was therefore led to adopt the view of Morgagni and Louis, and to infer that cyanosis is dependent either on obstruction at the pulmonic orifice, or on some other cause giving rise to venous congestion. This theory he regards as not only satisfactorily accounting for the discoloration of the skin and the dyspnœa and other symptoms; but he contends, that congestion of the venous system is always present when cyanosis exists, and is never found without the occurrence of cyanosis, unless there are satisfactory reasons for its absence.

M. Valleix,[2] one of the most recent writers on this subject, concludes "that cyanosis cannot be regarded as the pathognomonic symptom of communications between the

---

[1] On Cyanosis, or Morbus Cæruleus, by Moreton Stillé, M.D., Am. Jour. of Medical Sciences, N. S., vol. viii. 1844, p. 25.

[2] Guide du Médecin Praticien, 3mo éd., 1853, t. 1er, p. 738, and p. 96 of this work.

right and left cavities of the heart, and does not constitute a particular disease; but is common to several affections, and is only more or less frequent in each of them." He refers to a case under his care at the Hospice des Enfants Trouvés, in which an infant presented nothing unusual in its appearance, and went through the ordinary changes after birth, though the septum of the ventricles was a mere rudiment, and the freest intermixture of the currents of the blood must have existed.

All modern writers either adopt the exclusive views of M. Louis and Dr. Stillé—that cyanosis depends on venous stasis; or regard it as partly due to congestion of the venous system, and partly to the intermixture of the venous with the arterial blood. Those, however, who uphold the latter theory, differ widely in the extent to which they suppose the one or the other cause to operate. Dr. Walshe[1] says, " How is that doctrine (that cyanosis depends on congestion of the venous system) reconcilable with the fact that the most intense obstruction may occur in the adult without inducing cyanotic discoloration? How comes it, too, if communication between the two sides of the heart be so unimportant, that in 5 only, out of 71 cases of cyanosis collected by Stillé, was such communication wanting? Is it not likely that two things, so constantly found together, act as cause and effect, and that when a widely open foramen ovale has been found (as it certainly occasionally has) without previous cyanosis, some corrective condition, either organic or dynamic, has existed to prevent the intermixture? Doubtless constriction of the orifice of the pulmonary artery will increase the darkness of tint, by inducing venous stagnation; but I do not think there is evidence to show that, unassisted, such constriction can produce cyanosis." Dr. Speer,[2] who, in a recent number of the Medical Times and Gazette, has discussed this question, concurs in the views of Dr. Walshe.

---

[1] On Diseases of Lungs, Heart, and Aorta, 2nd. ed., 1854, p. 713.
[2] 1855, p. 412.

The frequency of the co-existence of cyanosis with communications between the two sides of the heart, is not overstated by Dr. Stillé. Of 124 cases of more or less marked blue discoloration of which I have collected notes or examined the specimens, in 112 there was either some defect in the partition of the ventricles, or patency of the foramen ovale, or a pervious arterial duct; and in only 12 cases does there appear to have been cyanotic discoloration, where the separation of the two currents of blood was complete. Small, however, as this number is, it is decisive against the theory of the intermixture of the currents of blood being the cause of cyanosis. In answer to the question of Dr. Walshe, it may be replied that the communication between the two sides of the heart is all-important in these cases ; for, in a large proportion of them, the obstruction to the transmission of the blood from the right to the left side is so extreme, that, without the existence of some abnormal passage, life could not be maintained even for the shortest period after birth.[1]

In the previous portion of this work I have alluded to various cases which bear out the inferences of Dr. Stillé. I have mentioned the case of a girl in whom there was an abnormal partition in the right ventricle, without any other malformation of the heart, and who was cyanotic during the several months she was under my observation, affording striking proof that cyanosis may exist without intermixture of the currents of blood. In the case of Dr. Hale, in which there existed only one ventricle; and in that of Dr. G. A. Rees, in which the pulmonary artery gave off the descending aorta, not the slightest lividity was observed ; so

---

[1] Dr. Stillé refers to five examples of the existence of cyanosis without any connexion between the two circulations. The most remarkable being that of Burnet, quoted by Bouillaud from the Jour. Univ. et Hebd. de Méd. et de Chir., t. i. 1830, in which the pulmonary artery was contracted ; and Marcet, Edin. Med. and Surg. Jour., vol. i. 1805, in which the lividity was due to chronic pulmonary disease. I may add references to the following, more or less closely resembling the case of Burnet :—Craigie, Tiedemann and Fohman, Chelius, Wilson, Breschet, Thore, Valleix, Hale, and Rees.

that these cases evince that the freest intermixture may exist without giving rise to cyanosis. Instances exhibiting the occurrence of cyanosis without intermixture, or the free entry of venous blood into the arterial system without cyanosis, are, however, much less frequent than cases which display a want of just relation between the intensity of the lividity and the amount of intermixture. In one of the instances of abnormal septum which I have mentioned as having fallen under my own notice, the aorta arose in great part from the right ventricle, so that a very large proportion of the blood circulating through the body must have been venous; yet there was no evidence that the boy had presented any material degree of cyanosis till shortly before his death. He had been an inmate of the Royal Free Hospital for an accident, about twelve months before, and nothing unusual was then observed in his appearance. The occurrence of cyanosis was apparently manifested after the pulmonary artery became the subject of disease, by which its capacity was still further diminished. In the case of Dr. Hess also in which the arteries were transposed, as the auricles both opened into the left ventricle and the right ventricle had no connexion with the corresponding auricle, but derived its blood from the left ventricle, the blood circulating in the systemic vessels must always have been, to a great extent, venous; yet the child was only livid during the paroxysms of asthma, and then not more so than often occurs in severe bronchitis. The absence of lividity in this case, on the supposition that cyanosis is dependent on congestion of the venous system, is readily explained by the orifices having been free from constriction, so that general venous congestion could not ordinarily have existed; but, on the theory that cyanosis is caused by the mixture of the currents of blood, its absence is quite inexplicable.

The fact that cyanosis is not always observed where abnormal communications exist between the two sides of the heart has been long known; and most pathologists have met with cases of imperfect closure of the foramen ovale or imperfection in the septum of the ventricles, in persons who had never presented appearances of cyanosis, or been suspected to have any defect in the conformation of the heart.[1] This has indeed been admitted by the supporters of the theory of intermixture, and various reasons have been assigned for the absence of cyanosis in such cases ; and especially it has been contended, that, provided the pressure on the two sides of the heart be equal, no intermixture will take place though either septum be defective. To this it may, however, be answered, that the cyanotic symptoms are by no means always congenital and sometimes do not appear till after many years, even when the freest intermixture of the currents of blood must have existed from birth. In cases of this kind, the accession of the cyanosis may generally be traced to the occurrence of more recent disease either in the heart or lungs, by which the original source of obstruction is aggravated. Thus, during an attack of rheumatism, inflammation may affect valves previously malformed, causing fibrinous exudations, and may so curtail still further the size of the opening into the pulmonary artery; or the aperture may be so rigid and unyielding, as not to expand sufficiently with the progress of growth; or an attack of bronchitis, by adding obstruction in the lungs to that which already existed in the heart, may cause the aggravation of cyanosis if previously present, or create it where it has not been before observed.

Lastly, cases frequently occur in which the variation in the degree of lividity cannot depend on any corresponding variation in the amount of intermixture. I recently saw an infant which suffered at intervals from the usual symptoms of malformation of the heart.[2] While quite quiet, there was no appearance of cyanosis, but paroxysms of dyspnœa with great lividity, were readily brought on by exposure to cold or by excitement or exertion. Under a mild alterative treatment, the paroxysms became less frequent, and had, indeed, ceased entirely, till the child, then three months old, took hooping cough attended with

---

[1] A case described in the Medical Gazette, 1843–44, by Dr. Mayo, affords a singular example of this kind, and one which is noticed by M. Durozier in the Comptes Rendus of the Société de Biologie, in which the foramen ovale was largely open in a female who died at the age of seventy-two, is even more remarkable.　　　　　　　　　　[2] See Case X., p. 91.

bronchitis, when they recurred with much greater severity, and the cyanosis became intense. On examination after death, the folds of the tricuspid valve were found somewhat adherent together, much thickened and indurated, and studded with recent fibrinous deposits; the right ventricle was hypertrophied and dilated and the pulmonary artery of large size. At the base of the septum of the ventricles there existed two apertures, leading from the left ventricle immediately below the origin of the aorta, into the sinus of the right ventricle. These apertures were larger on the left than on the right side, so that it was evident that the current of blood which had passed through them must have flowed from the left ventricle into the right; and, from their small size, they could neither have given passage to a large current, nor, from their hard and unyielding edges, could the quantity transmitted have been liable to material variation. The cyanosis could not, therefore, have been owing to the venous blood entering the left ventricle, and so being circulated through the body. Neither could the variations in its intensity have been due to any corresponding variation in the amount of intermixture. The different degrees of congestion of the venous system consequent on the increased difficulty in the transmission of blood through the lungs, could alone explain the recurrence of the paroxysms. In this case, also, the left arm and hand were at all times very livid and somewhat swollen; and it was found on examination, that the venous trunks on that side had been compressed by enlarged glands at the root of the lung and in their course.

From these considerations, we are, I think, warranted in inferring that the cyanosis is due to congestion of the venous system. I cannot, however, concur in the opinion of Laennec, that the lividity in cases of malformation differs in no degree from that which attends ordinary disease of the heart or lungs; and that in some forms of affection of the lungs, the discoloration of the skin is as considerable and as general as in cases of malformation. The cyanosis of malformation, when very marked, is much more intense than that from any other cause; but, occasionally, the lividity which attends pulmonary and cardiac disease is quite as great as in some instances of malformation. In support of this, I may mention the case of a boy, seventeen years of age, who was a patient of mine at the Royal Free Hospital in 1847, and had presented marked cyanosis from early life, yet in whom the discoloration was dependent on imperfect expansion of the lungs, connected with curvature of the spine, and the right ventricle was very greatly hypertrophied and dilated. A very remarkable example of lividity dependent on chronic disease of the lungs was published at the beginning of this century, by Dr. Marcet. In this instance, the body, that of a young woman of twenty-one years of age a patient at Guy's Hospital, was examined by Sir A. Cooper, and the evidence which it afforded of marked cyanosis, without intermixture of the venous and arterial blood, led Dr. Marcet to doubt the correctness of the then generally-received doctrine.[1] That in cases of pulmonary and ordinary cardiac disease, the cyanosis is usually so much less than where the heart is malformed, is probably to be ascribed to the amount of congestion being also less. In cases of acquired disease, were so small a proportion of blood submitted to the influence of the air and were the general congestion so extreme as in many instances of malformation, life could not be maintained. In acute affections, also, the integuments generally become more or less œdematous, so that the lividity is masked.[2]

Dr. Stillé's observations point too exclusively to contraction of the pulmonary orifice as giving rise to the congestion on which the cyanosis is dependent. In the previous pages, cases have been adduced in which the obstruction was caused by an abnormal septum in the right ventricle. An

---

[1] Edin. Med. and Surg. Jour., vol. i. 1805, p. 412.

[2] In some cases of chronic cerebral disease in children I have observed the patients to be very livid. The case before mentioned of atrophy of the brain and idiotcy was sent to me by my friend, Mr. Gay, who supposed, from the extreme lividity of the child, that it was probably a case of cardiac cyanosis; yet after death the heart was found quite natural.

instance has just been mentioned in which it was dependent on disease of the tricuspid valves; and, as before stated, it is sometimes caused by imperfect expansion of the lungs.

On the other hand, it has been shown that great contraction of the orifice of the pulmonary artery when it occurs in adult life, is not necessarily productive of cyanosis. The case of Dr. Hamilton Roe before mentioned, that of Dr. Ogle (Path. Trans. vol. v. p. 69), and one which is published in this work (Case XV. p. 122), equally prove that cyanosis is not always caused by even great *congenital* contraction of the pulmonary orifice. Various cases which have been published—I may especially refer to that of Dr. Craigie and others related in this work—also show that the cyanosis when present does not always bear a strict relation to the amount of obstruction. In all exceptional cases of this kind, however, I believe it will be found that the right ventricle has acquired such an increase of power, as to be able to overcome the difficulty in transmitting the blood through the contracted orifice, and so the occurrence of general congestion has been prevented. This " conservative hypertrophy" as it is termed by Dr. Ogle, is specially said to have existed in his case ; and in the instance just referred to the patient died of phthisis, and it is probable that the lividity had become less with the gradual diminution in the amount of blood circulating in the body with the progress of the pulmonary disease.

The inferences to be drawn from the facts brought forward appear to be, that, while obstruction to the flow of blood through the lungs or from or into the right ventricle, giving rise to general venous congestion, is the essential cause of cyanosis, the intensity of the lividity and its peculiar colour, are modified by other circumstances.

The chief of these modifying circumstances are, I believe, the following :—

1*st*. It is probably necessary to the production of intense cyanosis, that, as suggested by Dr. Chevers, the obstruction to the circulation should either have been present before birth, when the capillary vessels are naturally more capacious than in the adult; or, that it should have existed before the full development of the body was attained, and while the entire vascular system was more readily dilatable; or, at least, that it should have been of long duration, so that the capillary vessels may have become greatly expanded.

2ndly. The condition of the integuments also probably materially affects the production of cyanosis. In cases in which the peculiar blue or black colour is observed, the skin is usually very thin and transparent, and the body generally emaciated. Where, on the contrary, the discoloration is rather of a deep rose tint, the patients are not much emaciated, or are even in some cases tolerably well nourished; and, where the skin is pallid, there is either no material congestion, or it is masked by the œdematous condition of the integuments.

3rdly, and lastly. There can be no doubt that the intensity and peculiar tint of the cyanosis must be much affected by the colour of the blood in the vessels. Where a very small portion only can be submitted to the influence of the air in the lungs, the whole mass must be of an unusually deep colour, and the hue of the surface generally will be proportionately dark.

DURATION OF LIFE IN PERSONS LABOURING UNDER DIFFERENT FORMS OF MALFORMATION.

After the description which has been given of the various malformations, it will readily be understood that there is considerable difference in the influence they exert upon the duration of life.

Where there is only some slight irregularity in the development of the heart, so that small openings exist in the septa of the auricles or ventricles, the defect is of very little importance; indeed, it is by no means uncommon for such openings to be found in the hearts of persons who have died at advanced periods of life, and have never presented any signs of cardiac disease. On the other hand, where the arrest in the development of the heart is more extensive and is combined with some form of obstruction, it becomes a source of serious suffering and the duration of life is necessarily limited to a comparatively short period.

In those cases in which there is *moderate contraction of the orifice or trunk of the pulmonary artery, while the heart is otherwise well formed,* the increased power of the right ventricle may so far overcome the difficulty of transmitting the blood through the lungs, as to maintain the balance of the circulation, and allow a considerable amount of health and vigour to be enjoyed for many years. This was well illustrated by the case related at the Pathological Society, by Dr. Hamilton Roe, in which the patient lived to the age of thirty, and had been noted for his performances as a pedestrian. The patients whose cases are recorded by Tiedemann and Fohman, and by Chelius, were twenty-one and twenty-six years of age; and the subject of my own case attained the age of twenty-three. The diseased pulmonic valves, figured by Dr. Carswell, occurred in a man who died when forty years of age; and the patient who was under the care of Dr. Graham in the Edinburgh Infirmary, survived to the age of forty-four, and had been able to follow his employment of a navigator, working on one of the railways till six weeks before his death. In the case reported by M. Fallot, the subject attained to sixty-three years of age. In other instances, as in those of Burnet, Ogle, and Hicks, the patients died in early life. In this description of cases, however, the congenital origin of the disease is open to doubt.

Where *the foramen ovale is open,* the obstruction at the pulmonary orifice is generally greater than in the class of cases just mentioned, and the duration of life is therefore less. Instances, however, have occurred in which the patients have survived for many years. Of 20 persons presenting this form of anomaly of whose cases I possess notes, 11 are recorded to have lived to the age of fifteen years and upwards. Of these, 6 died at the ages of twenty,

twenty-eight, twenty-nine, thirty-four, forty, and fifty-seven ; and my own patient was twenty years of age. In 3 cases the ductus arteriosus also was open, and the subjects of the malformations attained the ages of ten months, fifteen months, and twenty-nine years.

Where, with *contraction of the pulmonary orifice the septum cordis is imperfect,* so that the aorta has a more or less direct communication with the right ventricle and the obstruction must have occurred at an early period of fœtal existence, the duration of life is still further curtailed: Of 64 such cases, 14 only survived the age of fifteen ; and of these, 1 died at the age of fifteen, 2 at sixteen, 1 at seventeen, 1 at eighteen, 1 at twenty, 2 at twenty-one, 1 at twenty-two, 1 at twenty-three, 3 at twenty-five, and 1 at thirty-nine years of age. In the five cases related in the previous portion of this essay, death took place at seventeen months, two years and five months, six years and a half, nine, and nineteen years. In cases of this description, the open state of the foramen ovale and the imperfection in the ventricular septum, so far from adding to the danger, really afford the means of relief to the overcharged right auricle and ventricle, without which life could not be prolonged for any considerable period. It does not, however, appear, that, provided the septum of the ventricles is very imperfect, the closure of the foramen ovale materially affects the prospects of longevity. The mean duration of life in the cases in which the septum of the ventricles was imperfect and the foramen ovale open, and in those in which apertures existed in the septum cordis and the foramen was closed, having been nearly equal. In only 8 of the 64 cases was the ductus arteriosus pervious, and in these, 1 of the patients died at the age of seventeen months, 1 at two years, 2 at three years, 1 at seven years, 1 at nine years and eleven months, 1 at thirteen years and a half, and 1 at nineteen years.

Where the *pulmonary artery is entirely impervious,* the usual duration of life is still less than in the latter class of

cases. Of 28 such instances, in which the patients survived
for a longer or shorter period after birth, 14 died before the
age of three mouths, 3 between three and six months, 4
between six and twelve months, 3 between twelve months
and two years, 3 lived to the ages of nine or ten, and 1 to
twelve years. Dr. Hare's patient attained the age of nine
months; and in the two cases here related the patients lived
nine days, and eleven months and two weeks. The duration
of life depends in cases of this kind upon the facility with
which the blood can be transmitted to the lungs; and is
therefore more likely to be prolonged when the septum of
the ventricles is imperfect or the foramen ovale open and
the ductus arteriosus pervious. But instances are on
record where the patients have survived for considerable
periods, though the lungs were dependent for their supply
upon enlarged bronchial arteries, or supplementary vessels
for the aorta or its branches. In the cases of this
kind recorded by Dr. Chambers and Dr. Babington, the
patients lived to the ages of nine to ten. In those of
Dr. Shearman and Dr. Crisp, in which the ductus arteriosus
probably transmitted the blood to the lungs, the children
lived to nine and twelve years of age. In the case reported
by Dr. Ramsbotham as one of absence of the pulmonary
artery, but in which that vessel existed though only in a
rudimentary condition, the patient died when sixteen years
old. In the case of Dr. Sibbald, in which there were
branches given off from the aorta and distributed to the lungs,
the child lived ten months; and in that of Dr. Buchanan,
where the lungs were also supplied direct from the aorta,
the child was six and a half months old. In both cases the
septum of the ventricles was imperfect.

Where the arrest of development is still more complete, so
that the *heart consists of only one ventricle with one or two
auricles,* the period for which the patients survive is usually
very limited: but it is very remarkable that four persons
who presented this condition obtained the ages of eleven,
sixteen, twenty-three, and twenty-four years. Dr. Hale's
patient lived nineteen weeks. Of five cases in which the

heart is said to have consisted only of a single auricle and ventricle, the infants survived seventy-eight hours, seventy-nine hours, seven days, and, in two cases, three days; but in four cases in which the arrest of development was less complete, the patients lived ten days, ten weeks, four months, and ten and a half months. In the instances of very complicated anomaly described by Drs. Buchanan, Kussmaul, and Dickenson, the patients were respectively four years, two and a quarter years, and three and a half years old.

I have not thought it necessary, in speaking of the age of patients presenting these extreme forms of malformation, to allude to the cases of Pozzi and Lanzoni, for, as previously stated, they are too imperfectly reported for their nature to be distinctly understood. Neither can I regard the statement here given of the duration of life in the published cases generally, as showing the ordinary viability of persons labouring under such malformations. The circumstance of any one surviving for several years with a heart which was found to present any of these conditions, would be regarded as so remarkable as to secure the case being placed on record; whereas it is probable that similar defects may have been frequently met with in infants without attracting much attention. The inference to be drawn from the facts is rather that the curtailment of life generally bears reference to the degree of impediment to the circulation of the blood; but that there is scarcely any amount of arrest of development which is not compatible with the occasional prolongation of life for some years.

Where *constriction exists at the commencement of the infundibular portion of the right ventricle,* the duration of life bears reference to the degree of obstruction which is occasioned, and to the extent of the coincident defects in the conformation of the heart—the effects corresponding indeed very closely with those which result from an impediment at the pulmonic orifice. Thus in the cases in which, in addition to the obstruction caused by the constriction, there was a communication between the two ventricles, the patients survived to the ages of nine, ten, twelve, fourteen, twenty, and twenty-two, and the subjects of my own cases of the kind died at the ages of seven and fifteen.   In these instances the pulmonary orifice also was contracted, though the impediment thus created was much less than that caused by the septum between the sinus and infundibular portion of the ventricle.   In Mr. Le Gros Clark's case the openings between the two parts of the ventricle were very small, but the pulmonic orifice was not contracted, and the boy lived to the age of nineteen.   In Dr. Thompson's case the septal obstruction was moderate, and the pulmonary orifice was enlarged, and the patient died at the age of thirty-eight. In Mr. Hutchinson's patient the obstruction was very great, and the heart was more imperfectly developed than in any of the others, yet she survived to the age of twelve.   In the case which I have related, in which the heart was naturally formed except for the existence of the septum, the patient died at the age of five years, of hæmorrhage from the throat or stomach during an attack of scarlatina; and M. Claude Bernard's patient, who presented a similar defect in a heart otherwise well formed, attained the age of fifty-six.   The congenital origin of the disease may, however, be disputed in this case; indeed, the obstruction is supposed by the author to have been the result of endocarditis, occurring during an attack of acute rheumatism.

Cases have before been mentioned showing that an open state of the foramen ovale and a pervious condition of the ductus arteriosus, when not associated with other serious defects in the conformation of the heart, are occasionally met with in persons who have attained middle or even advanced age, and who have never evinced marked signs of cardiac disturbance or defect.

The *transposition of the main arteries* appears to be a form of malformation which is incompatible with the maintenance of life for any considerable period after birth.   Of 21 cases of this description, 4 proved fatal within the first week, 1 in the second, 1 in the third week; 1 at the fourth

week ; 1 at the sixth week ; 2 at two months ; 2 at ten weeks ;
1 at five months ; 1 at twenty-two weeks ; 1 at seven months,
and 1 at ten months ; and 4 others at two years and six
months, two years and seven months, two years and eight
months, and two years and nine months. In the case of Mr.
Gamage, in which the viscera of the body generally were
transposed, the infant lived fifteen weeks. In that reported
by Dr. Hess and myself, in which there were other defects in
the conformation of the heart, the child survived eight months.
The existence of defect in the septum of the ventricles, though
of comparatively rare occurrence in this class of cases,
appears, as might be expected, to be favourable to the prolongation of life. Of 4 such cases, 3 of the patients survived
from seven months and eight days to two years and nine
months. The closure of the ductus arteriosus when the
septum of the ventricle is entire, would, on the contrary,
seem to add so greatly to the difficulty in transmitting the
blood to the lungs, as to be scarcely compatible with the
maintenance of life. In 4 cases, however, in which it is
stated that this condition existed and the only communication between the two sides of the heart was through the
foramen ovale, the patients died at the ages of fifteen weeks,
one year and six months, two years and six months, and
two years and eight months.

The duration of life in the form of malformation in which
*the aorta distal to the left subclavian artery is contracted or
impervious, and the descending aorta is wholly or chiefly supplied through the pulmonary artery,* is generally very limited.
The ready outlet for the blood from the right ventricle
through the large ductus arteriosus apparently prevents the
expansion of the lungs, and the children die with symptoms
of dyspnœa, syncope, and exhaustion. The infant whose
case is described by M. Pamard lived only thirty-six hours.
In the two cases which occurred to Sir Astley Cooper, the
infants lived two days, and eight months. In M. Gibert's
case the child survived twelve days, and Dr. Rees' patient
ten weeks. In the more aggravated form of this malformation described by Steidele, the infant died shortly

after birth, and the subjects of the cases related by Dr. Greig and Dr. Wale Hicks were still-born.

When, however, the contraction of the aorta is at the time of birth only slight, though the ductus arteriosus may remain pervious, the patient may survive to adult or middle age. Thus the subject of my own case of the kind was about thirty; and Rokitansky has reported instances in which the patients died at the ages of twenty-one, twenty-two, twenty-three, twenty-four, thirty-two, and forty-three. When, though the constriction at birth is only slight so that the duct may close, the contraction may ultimately become very great or the calibre of the vessel may be entirely obliterated, life may yet be prolonged to even very advanced age. From the paper to which I have previously referred it will be found, that in cases of only partial constriction, patients survived to the ages of fifty-four, sixty-nine, and ninety-two; and even when the aorta was entirely obliterated, the ages of forty-two, forty-five, fifty, and fifty-seven, were attained; though of course others, under both circumstances, died at much earlier periods of life.

When the heart is well formed, its *malposition* within the thorax is not necessarily productive of such serious inconvenience as to interfere with the duration of life. In Dr. Sampson's case the patient was thirty years of age; in Dr. Baillie's, nearly forty; and in M. Meckel's it would appear from the plate, that the subject of the malformation must have been an adult. Instances are, indeed, on record in which the heart with the other viscera of the body have been transposed, in persons who survived to advanced age and had never presented any evidence of embarrassment of the circulation. The subject of the case described by M. Méry had been a soldier, and died at seventy-two years of age; a patient mentioned by Riolan was eighty; and a still more remarkable instance is recorded by M. Bosc, in which the patient lived to the age of eighty-four. I have also referred to persons still living, at the ages of twenty-one, twenty-two, and forty-eight, who presented the same defect. When the heart is situated in the abdomen, life also may be much prolonged. The man mentioned by Deschamps, in whom the heart occupied the situation of the left kidney, was a soldier who had served for many years in the army. When, on the other hand, the heart occupies a position entirely external to the cavity of the chest, there is most generally some serious defect in the conformation of the organ or other viscera of the body, so that life is usually maintained only for a very limited period. In Mr. Sidney Jones' case, in which the heart was situated entirely external to the thorax, the infant only lived thirteen hours; and in the case mentioned by Dr. O'Bryan, in which the organ was situated in part externally, the child survived three months. In the cases of MM. Follin, Cruveilhier and Monod, and Mr. Daniel, the children only lived a few hours.

The *absence of the pericardium* does not appear to affect the functions of the heart or to interfere with the duration of life, though the organ, under such circumstances, very generally contracts adhesions to the adjacent parts. All the cases quoted, occurred in persons who had attained adult or middle age, and only one or two of them were known to have manifested during life any symptoms of disorder of the heart or circulation.

The *malformations of the valves* vary in their effects according to the nature of the irregularity. When the number of the segments is defective and especially when the whole of them are united together, some obstruction is generally occasioned, the valves are apt to become the seat of subsequent disease, and life is usually more or less curtailed. I have, however, seen only two valves at the aortic orifice occasionally in persons who have died in advanced life and frequently in middle age, without the segments having presented any appearances of more recent disease; and similar facts have been noted by others. When the number of the segments is in excess, no inconvenience appears, as previously mentioned, to result, and life may be maintained to the full period. I have figured a specimen in which there were four valves to the pulmonary artery, taken from a woman seventy-five years of age, in whom the segments were free from disease, and no functional disturbance had apparently been occasioned.

The causes of death in cases of malformation of the heart are—

1st. Cerebral disturbance resulting from the defective aëration of the blood and congestion of the brain.

2ndly. Imperfect expansion, collapse and engorgement of the lungs.

3rdly. Effusion into the cellular tissue and serous sacs, from failure of the power of the heart, or recent disease superinduced on the original cardiac defect.

4thly. Exhaustion from the imperfect performance of the respiratory functions and the circulation of blood in great part venous.

5thly. Other diseases predisposed to by the defective conformation of the heart; as apoplexy or paralysis from engorgement or softening of the brain, or extravasation of blood; epistaxis; congestion or inflammation of the lungs, croup, bronchitis, pneumonia, pulmonary apoplexy and hæmoptysis, &c.; disorder of the digestive organs, vomiting, diarrhœa, jaundice, &c.; renal affections, &c.

6thly. Other diseases occurring accidentally, or not necessarily connected with the defective conformation of the heart. Peri-, endo- and myo-carditis; tuberculous affections of the lungs or other viscera, &c.; embolism of orifices of the heart and of the vessels, especially the pulmonary artery; pleurisy; &c.

Of these different causes the two first are by far the most frequent; and especially in children in which the heart is very imperfect, and life is therefore only prolonged for a very limited period. Serious dropsical affections less frequently occur than the degree of obstruction to the circulation would lead us to expect. Gradual exhaustion from imperfect nutrition and disorder of the digestive organs, are occasionally the causes of death both in infants and older subjects.

I have already mentioned that cerebral and pulmonary

diseases and hæmoptysis are frequently fatal when the patients survive several years, and death also occasionally occurs from tuberculous affections of the lungs, especially when life is maintained to the age of ten or fifteen.   In a patient of my own, who lived to the age of twenty,[1] in whom the pulmonary artery was greatly contracted and the foramen ovale largely open, and in a case related by Dr. Leared,[2] in which the patient died at the age of eight and presented similar defects, the lungs were found tuberculous. In the young woman whose case was reported by Dr. Rams-botham,[3] as one of absence of the pulmonary artery with compensatory branches from the aorta but in whom the vessel was only very small, death occurred from phthisis at the age of sixteen ; and that disease was also partly the cause of death in the boy whose case is related by Dr. Shearman,[4] in whom the pulmonary artery was obliterated and who died at the age of nine.   Death ensued from the same cause in another boy, thirteen and half years old, mentioned by Dr. Cheevers[5] of Boston, in whom the pulmonary artery was nearly impervious, the septum of the ventricles imperfect, and the foramen ovale and ductus arteriosus open.   In cases related by Lexis,[6] Huss,[7] Gregory,[8] Bertody and Dunglisson,[9] Gintrac,[10] and Louis,[11] in persons, aged respectively five years and nine months, six years, eighteen, twenty-one, twenty-one, and twenty-five years, in whom the pulmonary artery was contracted and the septum of the ventricles defective, the lungs were also tuberculous.

---

[1] Case XIII. p. 112 supra.   [2] Dublin Journal, N. S., vol. x. 1850, p. 223.
[3] Med. and Phys. Jour., vol. lxi. (N. S., vol. vi.), p. 548. In a female sixteen years of age.
[4] Prov. Med. and Surg. Jour., 1845, p. 484.   In a female nine years of age.
[5] New England Journ. of Med. and Surg., vol. x. 1821 (N. S., vol. v.) p. 217. In a boy thirteen and a half years of age.
[6] Lancet, 1835–36, vol. ii. p. 433, and referred to at p. 58 supra.
[7] Gaz. Méd. de Paris, 2me série, t. xime an. 1843. p. 91, referred to at p. 60 supra.
[8] Med. Chir. Trans., vol. xi. 1820, p. 296.
[9] Phil. Med. Ex., 1845, quoted in Dublin Jour., vol. xxviii. 1845, p. 300.
[10] Sur la Cyanose, obs. 45.
[11] Arch. Gén. de Méd., 2me série, t. iii. 1823, obs. 9, and Mémoires et Re-cherches Anatomico-pathologiques, 1826, obs. 10, p. 313.

I have been the more particular in alluding to these cases as they are opposed to the assertion of Rokitansky that tuberculosis does not "coexist with congenital vices of formation of the heart, or great arterial trunks, which, with their complications, result in venosity and cyanosis,"[1] and that "all cyanoses, or rather all forms of disease of the heart, vessels or lungs, inducing cyanosis of various kinds and degrees, are incompatible with tuberculosis, against which cyanosis affords a complete protection."[2]

In the cases which I have referred to, the cyanosis existed to a marked degree in six, but was only occasionally noticed in three. In most of them, it was observed at or shortly after birth and continued throughout life. In all the cases but a small portion of blood could have been submitted to the influence of the air in the lungs, the freest intermixture of the currents must have existed, and the blood circulating in the systemic vessels must have been to a great extent venous. Yet this condition did not prevent the occurrence of tuberculosis. On the contrary, active tuberculous disease was present in every case, there being either miliary tubercles, softened tubercle, or cavities in the lungs. As these cases constituted nine, or 16·07 per cent. of fifty-six, in which the patients with different forms of malformation survived the age of eight, it might even be supposed that tuberculous affections are more common in persons with defects in the conformation of the heart than in the population at large.[3] This would, however, probably be carrying the inference too far. It is possible that the venous condition may, as supposed by Laennec, be in some degree opposed to the occurrence of tuberculous affections; but this opposition certainly in no degree amounts to an incompatibility, as asserted by Rokitansky. Every medical man of much experience has also met with cases in

---

[1] Pathological Anatomy, Sydenham Society's translation, vol. i. p. 316.

[2] Pathological Anatomy, Sydenham Society's translation, vol. iv. p. 251.

[3] From the Registrar-General's Report for 1854, it appears that the cases of consumption constituted only 9·1 per cent. of the total number of deaths in England and Wales.

which tuberculosis has occurred in persons long subject to chronic bronchitis and asthma. In the case of malformation related by MM. Aran and Deguise,[1] the patient died of disease of the hip.

Since the publication of the former edition of this work, my attention has been drawn to various other cases which have either been recorded more recently or previously ; and which concur in showing that the generalization of Rokitansky, as to the incompatibility of the tuberculous affections with cyanosis or a venous condition of the blood, was too hastily advanced. Thus, a case has been related by M. Barth,[2] in which a patient who had been very livid died of bronchitis and pleurisy, and an aperture existed in the septum of the ventricle; the lungs were found to be tuberculous. The case of M. Gubler[3] may also be referred to as having presented conditions which might have given rise to cyanosis, in which the lungs were tuberculous ; and cases described by Drs. Markham,[4] and Wilks,[5] and Mr. Nunnelcy,[6] equally illustrate the coincidence of the two conditions. In a case recorded by Dr. Hillier,[7] a highly cyanotic child in whom there was a small aperture in the septum of the ventricles had tubercle in the mesenteric glands. I have also met with two cases[9] in which persons labouring under malformation of the heart died of phthisis. In one of them there was a widely open foramen ovale ; and in the other the orifice of the pulmonary artery was greatly contracted by adhesion of the valves, although there was no connexion between the two sides of the heart.

DIAGNOSIS.

The detection of the existence of malformation of the heart, in ordinary cases, when the patient is seen in early

[1] Lancet, 1844 ; and Bullet. de la Soc. Anat., 1842, &c.
[2] Bullet. de la Soc. Anat. de Paris, 3me série, t. i. année 10, 1835, p. 145.
[3] P. 59 supra.   [4] Path. Trans., vol. xi. 1859–60, p. 68.
[5] Ibid., vol. x. 1853–59, p. 79.   [6] Ibid., vol. xiii. 1861–62, p. 42.
[7] Ibid., vol. xii. 1860–61, p. 76.   [8] Cases XIV. p. 116; and XV. p. 122.

life, can scarcely present any difficulty. The statement that palpitation, dyspnœa, and more or less cyanosis, had existed since birth or shortly after, and the evidences of obstructed circulation at the time of examination, render the case sufficiently clear. M. Louis, indeed, regards the occurrence of "suffocative attacks brought on by the slightest cause, often periodic, and always very frequent, and accompanied or followed by syncope, and with or without the blue discoloration of the body generally," as pathognomonic of communications between the right and left cavities of the heart; and the cyanotic discoloration when present can scarcely be mistaken. But the ordinary symptoms may be absent or may exist only to a slight degree; or the patient may not be seen till after he has attained the age of puberty or manhood, and there may be no satisfactory history of his previous state of health to aid the diagnosis. Though, in cases of this kind, if the patients had been under medical care, it is quite possible that sufficiently characteristic signs might have been observed, we are sometimes assured by the patient and his friends that he had enjoyed good health, had been capable of following a laborious occupation, and had presented nothing unusual in his appearance, until shortly before the time at which he falls under our notice. In such cases, then, it may be extremely difficult to decide whether the patient labours under some kind of malformation, or under ordinary disease of the heart; and the differential diagnosis can only be effected by a careful examination and analysis of the general symptoms and physical signs.

In all cases, also, the detection of the precise form of malformation must be a task of considerable difficulty, and in some instances entirely impracticable. Where an infant suffers from great difficulty of breathing and palpitation, and is intensely and constantly cyanosed, at or immediately after birth, it may be inferred that it labours under some serious malformation occasioning great obstruction to the circulation of the blood, as obliteration or great contraction of the pulmonic orifice, or transposition of the aorta and pulmonary artery. On the contrary, when the symptoms do not manifest themselves at so early a period, and are less constant and intense, there is probably only some slighter malformation, as a moderate amount of contraction at the pulmonary orifice. Of 181 cases of various forms of decided and important malformation, of which I have collected notes, in 90 there existed more or less contraction of the orifice of the pulmonary artery or other sources of obstruction to the exit of the blood from the right ventricle; and in 29 others the orifice or trunk of the vessel was obliterated. In those patients who survived the age of twelve, the entrance of the blood into the pulmonary artery was interfered with in a much larger proportion of cases, or in 38 out of 45. So that, in any given case of malformation, especially after the age of fifteen, the probability is that the pulmonary artery is contracted. If this be the case, a loud systolic murmur will be heard in the præcordial region, and most intensely at the level of the nipple and between that body and the sternum. It will be audible very distinctly in the course of the pulmonary artery, or from the base of the heart towards the middle of the left clavicle; and less distinctly in the course of the aorta, or at the upper part and right side of the sternum. If the pulmonic orifice be permanently open as is often the case, especially where the whole of the valves are united, there may also be a diastolic murmur; but, from the very small size of the aperture in most instances, the regurgitant current is probably generally too slight to generate a distinct murmur. Most usually with considerable contraction of the pulmonary orifice, the septum of the ventricles is defective, and the aorta derives its supply of blood from both ventricles; and, if so, a systolic murmur may probably be produced by the meeting of the two columns of blood in the ascending aorta, which may modify the signs observed. Generally, in such cases, the aorta is unusually large, and, from the powerful reaction on the valves during the diastole of the heart, a loud ringing second sound is heard on listening at the upper part of the sternum. With these signs there will be perceived those of

o

hypertrophy and dilatation of the right ventricle and auricle, and frequently a distinct jugular pulsation will be observed. The heart being much increased in size and its walls hypertrophied, the dull space will be extended beyond its usual limits, especially towards the right side. From the yielding of the parietes in early life, the præcordia is also generally prominent. The impulse of the heart is usually powerful; and frequently a distinct purring tremor may be felt over the situation of the pulmonic orifice. The pulse is generally quick, small and weak. It has been thought that, in cases of obstruction at the right side of the heart, the patient is disposed to let the head hang down so as to compress the chest, rather than to adopt the upright position, which we most frequently see selected by patients with disease of the left orifices; but I have seen patients with aortic disease hang themselves completely over the side of the bed, so that this rule does not certainly apply.

If the evidence of obstruction at the pulmonic orifice be tolerably conclusive, we may safely infer there is either a deficiency in the septum of the ventricles or a patent foramen ovale; for one or other of these defects almost invariably co-exists with that condition. An aperture in the septum of the ventricles, without other malformation, would probably be attended by a murmur, caused by the flow of blood through the abnormal opening from the left ventricle into the right ventricle or auricle. The detection, therefore, of a systolic murmur at the base of the heart, without signs of obstruction at the aortic or pulmonic orifice, might lead to a suspicion that such a communication existed. This surmise would be strengthened if the murmur were not propagated in the course of the pulmonary artery or aorta; and especially if the patient were long under notice, and constantly presented the sign, without other evidence of cardiac disease or defect; and without having had any disease or accident during life, which could probably have produced such a change in the heart as would be likely to be attended by a permanent murmur.

I do not know that there are any means of detecting with

tolerable certainty the open state of the foramen ovale.[1] There are also other malformations of the heart, such as transposition of the aorta and pulmonary artery, which could not be at all diagnosed during life.

In some cases, as where the ductus arteriosus or foramen ovale remains open, the diagnosis may be aided by ascertaining whether the infant has been born prematurely or at the full period.

It has already been mentioned that the malformations of the valves do not necessarily entail any interference with the functions of the heart. When they lead to disease of the organ, the symptoms and physical signs will be those of valvular disease dependent on any other cause. I believe that when a patient in early or adult life labours under symptoms of valvular disease, more especially at the aortic orifice, without having previously sustained any severe injury or strain, and without having had any serious rheumatic attack or other obvious cause to which the symptoms can be ascribed, we shall generally be correct in inferring that the valves are malformed.

MEDICAL TREATMENT.

The medical treatment in cases of malformation of the heart must consist, first, in the hygienic management of the patient, so as to maintain the circulation, and give tone to the general system ; secondly, in the avoidance of the various

---

[1] In the Pathological Transactions for 1856-57, Dr. Markham has described the case of an infant four years old who had been ill for three weeks and was slightly cyanotic, and died ten days after being first seen. In this case "a rough, loud systolic bruit was audible all along the base of the heart and in the whole of the left subclavicular region. It was distinctly heard below the nipple, and was scarcely audible at the heart's apex ; its point of greatest intensity was to the left of the upper part of the sternum. It was not audible at the right edge of the sternum in the course of the aorta. Afterwards the murmur was heard more loudly and at the upper part of the sternum and along its right border, and was remarkably loud in the whole of the upper part of the interscapular space." In this case the heart was in all respects healthy, except that the foramen ovale was so largely open as to permit the passage of the point of the finger, though partially covered by a loose membrane ; and as no other cause was detected for the production of the murmur than the passage of blood through the opening, Dr. Markham ascribed the abnormal sound to that cause.—Vol. viii. p. 142.

causes which may aggravate the existing defects in the conformation of the heart, or give rise to secondary disease in
other organs; and, thirdly, in the relief of the urgent
symptoms when they arise.

1st. The surface should be carefully protected against
cold, so as to economize the scanty power of generating
heat, and this may be accomplished by warm clothing, and
especially by wearing flannel, or some other woollen material,
or silk, next the skin. The patient should reside in a warm
situation if practicable, and the room which he occupies
should be maintained at an agreeable and equable temperature. His strength should be upheld by nutritious food;
and a moderate quantity of stimulus, malt liquor or wine,
should be allowed. The secretions should be regulated by
gentle exercise, in the nurse's arms if the patient be an infant, or in a carriage if an older person; and when the
weather is suitable he should frequently be in the open air.
The cutaneous functions, which may greatly compensate for
the defective aëration of the blood in the lungs, should be
promoted by warm or tepid bathing or ablution, followed by
friction of the surface.

2ndly. The patient should be kept quiet and undue excitement of mind and fatigue of body should be avoided. Sudden
exposure to cold and damp should be specially guarded
against, lest the patient should become the subject of any
form of pulmonary disease, which would greatly aggravate
the general congestion; or of rheumatism, which would
almost certainly implicate the heart. Care should also be
exercised that the food which is taken should be easily
digestible, and that the stomach should not be overloaded
or the secretions checked; as neglect of these circumstances
would be very likely to bring on attacks of dyspnœa or convulsions. Over-excitement of the brain should be especially
guarded against.

3rdly. For the relief of the paroxysms when they occur,
the first indication is to ascertain the cause by which they
have been induced. If undue exertion of body or excitement of mind have caused the attack, it will probably sub-

side on entire rest. If the stomach have been overloaded or indigestible food have been taken, an emetic or a mild aperient, may be given ; and if the symptoms do not afterwards subside, the patient may be treated by antispasmodics —the warm bath, ether, ammonia, &c. and anodynes.

In most cases of malformation the patients labour under dyspeptic symptoms, the bowels are generally torpid, and the attacks are often occasioned or aggravated by flatulency. For the relief of these symptoms small doses of hydrargyrum cum cretâ, with rhubarb, soda or magnesia, and followed by bitter tonics, are very beneficial. When there is much palpitation or pain in the region of the heart, hydrocyanic acid, hyoscyamus, and opiates are applicable. The Dover's powder is especially useful, not only by acting as a mild anodyne, but also by promoting diaphorèsis. The convulsive attacks to which cyanotic persons are subject, are often relieved by the application of a few leeches to the temples or behind the ears ; and in children which are teething the pressure of the gums may be removed by lancing. From, however, the scanty aëration of the blood and its consequent deficiency in fibrine, and from the excessive congestion of all parts of the system, the flow of blood either from leech-bites or incisions is apt to be excessive, and in some cases I have seen dangerous hæmorrhage. Great care should therefore be practised in having recourse to these means, that the loss of blood shall not exceed the required amount.

By the course of management which has been mentioned, in the slighter forms of malformation life may occasionally be prolonged for several years, and in some instances a considerable amount of constitutional vigour may be enjoyed ; but such cases are exceptional and only occur under the most favourable circumstances. In the more serious deviations from the natural conformation of the heart, which constitute the majority of the cases, the maintenance of life is unavoidably limited to a few days, weeks, or months, and the benefit to be derived from medicine is confined to affording some alleviation to the sufferings of the patient.

INDEX.

INDEX TO THE AUTHORS OF CASES QUOTED.

THE END.

London, New Burlington Street,
November, 1866.

MESSRS. CHURCHILL & SONS'

Publications,

IN

MEDICINE

AND THE VARIOUS BRANCHES OF

NATURAL SCIENCE.

A CLASSIFIED INDEX

TO

MESSRS. CHURCHILL & SONS' CATALOGUE.

MEDICINE—continued.

PAGE

Aldis's Hospital Practice 6
Anderson (Andrew) on Fever.. 7
Do. (Thos.) on Yellow Fever 7
Austin on Paralysis 7
Barclay on Medical Diagnosis.. 8
Do. on Gout.. 8
Barlow's Practice of Medicine 8
Basham on Dropsy 8
Brinton on Stomach 11
Do. on Ulcer of do. 11
Budd on the Liver 11
Do. on Stomach 11
Camplin on Diabetes.. 12
Chambers on Digestion 12
Do. Lectures 12
Cockle on Cancer 13
Davey'sGanglionic Nervous Syst. 15
Day's Clinical Histories 15
Eyre on Stomach 15
Fuller on Rheumatism 17
Gairdner on Gout 17
Gibb on Throat 17
Granville on Sudden Death .. 18
Griffith on the Skin 18
Gully's Simple Treatment .. 18
Habershon on the Abdomen .. 18
Do. on Mercury 18
Hall (Marshall) on Apnœa .. 18
Do. Observations .. 18
Headland—Action of Medicines 19
Hooper's Physician's Vade-Mecum 18
Inman's New Theory 22
Do. Myalgia.. 22
James on Laryngoscope 22
Maclachlan on Advanced Life.. 25
MacLeod on Acholic Diseases.. 25
Marcet on Chronic Alcoholism.. 25
Macpherson on Cholera 26
Markham on Bleeding 26
Meryon on Paralysis 27
Mushet on Apoplexy.. 27
Nicholson on Yellow Fever .. 27
Parkin on Cholera 28
Pavy on Diabetes 28
Peet's Principles and Practice of Medicine 28
Roberts on Palsy 30
Robertson on Gout 30
Sansom on Cholera 31
Savory's Compendium 31
Semple on Cough 32
Seymour on Dropsy 32
Shaw's Remembrancer 32
Shrimpton on Cholera 32
Smee on Debility 32
Thomas' Practice of Physic .. 35
Thudichum on Gall Stones .. 35
Todd's Clinical Lectures 36
Tweedie on Continued Fevers 36
Walker on Diphtheria 36
What to Observe at the Bedside 25
Williams' Principles 38
Wright on Headaches 39

MICROSCOPE.

Beale on Microscope in Medicine 8
Carpenter on Microscope 12
Schacht on do. 31

MISCELLANEOUS.

Acton on Prostitution 6
Barclay's Medical Errors 8
Barker & Edwards' Photographs 8

MISCELLANEOUS—cont⁴.

PAGE

Bascome on Epidemics 8
Blaine's Veterinary Art 10
Bourguignon on the Cattle Plague 10
Bryce on Sebastopol 11
Buckle's Hospital Statistics .. 11
Cooley's Cyclopædia 13
Gordon on China 17
Graves' Physiology and Medicine 17
Guy's Hospital Reports 17
Harrison on Lead in Water .. 19
Hingeston's Topics of the Day .. 20
Howe on Epidemics 21
Lane's Hydropathy 23
Lee on Homœop. and Hydrop. 24
London Hospital Reports.. ..25
Marcet on Food 25
Massy on Recruits 26
Mayne's Medical Vocabulary .. 26
Part's Case Book 28
Redwood's Supplement to Pharmacopœin 30
Ryan on Infanticide 31
St. George's Hospital Reports .. 31
Simm's Winter in Paris 32
Snow on Chloroform.. 33
Steggall's Medical Manual .. 34
Do. Gregory's Conspectus 34
Do. Celsus.. 34
Waring's Tropical Resident at Home 37
Whitehead on Transmission .. 38

NERVOUS DISORDERS AND INDIGESTION.

Althaus on Epilepsy, Hysteria 7
Birch on Constipation 10
Carter on Hysteria 12
Downing on Neuralgia 15
Hunt on Heartburn 21
Jones (Handfield) on Functional Nervous Disorders.. 22
Leared on Imperfect Digestion 23
Lobb on Nervous Affections .. 24
Radcliffe on Epilepsy 29
Reynolds on the Brain 30
Do. on Epilepsy 30
Rowe on Nervous Diseases .. 31
Sieveking on Epilepsy 32
Turnbull on Stomach 36

OBSTETRICS.

Barnes on Placenta Prævia .. 8
Hodges on Puerperal Convulsions 20
Lee's Clinical Midwifery 24
Do. Consultations 24
Leishman's Mechanism of Parturition 24
Pretty's Aids during Labour .. 29
Priestley on Gravid Uterus .. 29
Ramsbotham's Obstetrics 29
Do. Midwifery.. .. 30
Sinclair & Johnston's Midwifery 32
Smellie's Obstetric Plates.. .. 33
Smith's Manual of Obstetrics .. 33
Swayne's Aphorisms.. 34
Waller's Midwifery 37

OPHTHALMOLOGY.

Cooper on Injuries of Eye .. 13
Do. on Near Sight 13
Dalrymple on Eye 14

OPHTHALMOLOGY—cont⁴.

PAGE

Dixon on the Eye 15
Hogg on Ophthalmoscope .. 20
Hulke on the Ophthalmoscope 21
Jago on Entoptics 22
Jones' Ophthalmic Medicine .. 23
Do. Defects of Sight .. 23
Do. Eye and Ear 23
Macnamara on the Eye 25
Nunneley on the Organs of Vision 27
Solomon on Glaucoma 33
Walton on the Eye 37
Wells on Spectacles 37

PHYSIOLOGY.

Carpenter's Human 12
Do. Manual 12
Heale on Vital Causes 19
Richardson on Coagulation .. 30
Shea's Animal Physiology 32
Virchow's (ed. by Chance) Cellular Pathology 12

PSYCHOLOGY.

Arlidge on the State of Lunacy 7
Bucknill and Tuke's Psychological Medicine 12
Conolly on Asylums 13
Davey on Nature of Insanity .. 15
Dunn's Physiological Psychology 15
Hood on Criminal Lunatics .. 21
Millingen on Treatment of Insane 26
Murray on Emotional Diseases 27
Noble on Mind 27
Sankey on Mental Diseases .. 31
Williams (J. H.) Unsoundness of Mind 38

PULMONARY and CHEST DISEASES, &c.

Alison on Pulmonary Consumption 6
Barker on the Lungs 8
Billing on Lungs and Heart .. 10
Bright on the Chest 11
Cotton on Consumption 14
Do. on Stethoscope .. 14
Davies on Lungs and Heart .. 15
Dobell on the Chest 15
Do. on Tuberculosis .. 15
Do. on Winter Cough .. 15
Fenwick on Consumption.. .. 16
Fuller on Chest 17
Do. on Heart 17
Jones (Jas.) on Consumption.. 22
Laennec on Auscultation 23
Markham on Heart 26
Peacock on the Heart 28
Richardson on Consumption .. 30
Salter on Asthma 31
Skoda on Auscultation 26
Thompson on Consumption .. 35
Timms on Consumption 35
Turnbull on Consumption .. 37
Waters on Emphysema 37
Weber on Auscultation 37

TO BE COMPLETED IN TWELVE PARTS, 4TO., AT 7s. 6d. PER PART.

PART I. NOW READY.

A DESCRIPTIVE TREATISE

ON THE

NERVOUS SYSTEM OF MAN,

WITH THE MANNER OF DISSECTING IT.

By LUDOVIC HIRSCHFELD,

DOCTOR OF MEDICINE OF THE UNIVERSITIES OF PARIS AND WARSAW, PROFESSOR OF ANATOMY TO THE FACULTY OF MEDICINE OF WARSAW;

Edited in English (from the French Edition of 1866)

By ALEXANDER MASON MACDOUGAL, F.R.C.S.,

WITH

AN ATLAS OF ARTISTICALLY-COLOURED ILLUSTRATIONS,

Embracing the Anatomy of the entire Cerebro-Spinal and Sympathetic Nervous Centres and Distributions in their accurate relations with all the important Constituent Parts of the Human Economy, and embodied in a series of 56 Single and 9 Double Plates, comprising 197 Illustrations,

Designed from Dissections prepared by the Author, and Drawn on Stone by

J. B. LÉVEILLÉ.

MR. ACTON, M.R.C.S.

I.

A PRACTICAL TREATISE ON DISEASES OF THE URINARY
AND GENERATIVE ORGANS IN BOTH SEXES. Third Edition. 8vo. cloth,
£1. 1s. With Plates, £1. 11s. 6d. The Plates alone, limp cloth, 10s. 6d.

II.

THE FUNCTIONS AND DISORDERS OF THE REPRODUC-
TIVE ORGANS IN CHILDHOOD, YOUTH, ADULT AGE, AND ADVANCED
LIFE, considered in their Physiological, Social, and Moral Relations. Fourth Edition.
8vo. cloth, 10s. 6d.

III.

PROSTITUTION : Considered in its Moral, Social, and Sanitary Bearings,
with a View to its Amelioration and Regulation. 8vo. cloth, 10s. 6d.

DR. ADAMS, A.M.

A TREATISE ON RHEUMATIC GOUT; OR, CHRONIC
RHEUMATIC ARTHRITIS. 8vo. cloth, with a Quarto Atlas of Plates, 21s.

MR. WILLIAM ADAMS, F.R.C.S.

I.

ON THE PATHOLOGY AND TREATMENT OF LATERAL
AND OTHER FORMS OF CURVATURE OF THE SPINE. With Plates.
8vo. cloth, 10s. 6d.

II.

ON THE REPARATIVE PROCESS IN HUMAN TENDONS
AFTER SUBCUTANEOUS DIVISION FOR THE CURE OF DEFORMITIES.
With Plates. 8vo. cloth, 6s.

III.

SKETCH OF THE PRINCIPLES AND PRACTICE OF
SUBCUTANEOUS SURGERY. 8vo. cloth, 2s. 6d.

DR. WILLIAM ADDISON, F.R.S.

I.

CELL THERAPEUTICS. 8vo. cloth, 4s.

II.

ON HEALTHY AND DISEASED STRUCTURE, AND THE TRUE
PRINCIPLES OF TREATMENT FOR THE CURE OF DISEASE, ESPECIALLY CONSUMPTION
AND SCROFULA, founded on MICROSCOPICAL ANALYSIS. 8vo. cloth, 12s.

DR. ALDIS.

AN INTRODUCTION TO HOSPITAL PRACTICE IN VARIOUS
COMPLAINTS; with Remarks on their Pathology and Treatment. 8vo. cloth, 5s. 6d.

DR. SOMERVILLE SCOTT ALISON, M.D.EDIN., F.R.C.P.

THE PHYSICAL EXAMINATION OF THE CHEST IN PUL-
MONARY CONSUMPTION, AND ITS INTERCURRENT DISEASES. With
Engravings. 8vo. cloth, 12s.

DR. ALTHAUS, M.D., M.R.C.P.

ON EPILEPSY, HYSTERIA, AND ATAXY. Cr. 8vo. cloth, 4s.

THE ANATOMICAL REMEMBRANCER; OR, COMPLETE
POCKET ANATOMIST. Sixth Edition, carefully Revised. 32mo. cloth, 3s. 6d.

DR. McCALL ANDERSON, M.D.

I.
PARASITIC AFFECTIONS OF THE SKIN. With Engravings.
8vo. cloth, 5s.   II.
ECZEMA. 8vo. cloth, 5s.

III.
PSORIASIS AND LEPRA. With Chromo-lithograph. 8vo. cloth, 5s.

DR. ANDREW ANDERSON, M.D.

TEN LECTURES INTRODUCTORY TO THE STUDY OF FEVER.
Post 8vo. cloth, 5s.

DR. THOMAS ANDERSON, M.D.

HANDBOOK FOR YELLOW FEVER: ITS PATHOLOGY AND
TREATMENT. To which is added a brief History of Cholera, and a method of Cure.
Fcap. 8vo. cloth, 3s.

DR. ARLIDGE.

ON THE STATE OF LUNACY AND THE LEGAL PROVISION
FOR THE INSANE; with Observations on the Construction and Organisation of
Asylums. 8vo. cloth, 7s.

DR. ALEXANDER ARMSTRONG, R.N.

OBSERVATIONS ON NAVAL HYGIENE AND SCURVY.
More particularly as the latter appeared during a Polar Voyage. 8vo. cloth, 5s.

MR. T. J. ASHTON.

I.
ON THE DISEASES, INJURIES, AND MALFORMATIONS
OF THE RECTUM AND ANUS. Fourth Edition. 8vo. cloth, 8s.

II.
PROLAPSUS, FISTULA IN ANO, AND HÆMORRHOIDAL
AFFECTIONS; their Pathology and Treatment. Second Edition. Post 8vo. cloth, 2s. 6d

MR. W. B. ASPINALL.

SAN REMO AS A WINTER RESIDENCE. With Coloured Plates.
Foolscap 8vo. cloth, 4s. 6d.

MR. THOS. J. AUSTIN, M.R.C.S.ENG.

A PRACTICAL ACCOUNT OF GENERAL PARALYSIS:
Its Mental and Physical Symptoms, Statistics, Causes, Seat, and Treatment. 8vo. cloth, 6s.

DR. THOMAS BALLARD, M.D.

A NEW AND RATIONAL EXPLANATION OF THE DIS-
EASES PECULIAR TO INFANTS AND MOTHERS; with obvious Suggestions
for their Prevention and Cure. Post 8vo. cloth, 4s. 6d.

DR. BARCLAY.

I.

A MANUAL OF MEDICAL DIAGNOSIS. Second Edition.
Foolscap 8vo. cloth, 8s. 6d.

II.

MEDICAL ERRORS.—Fallacies connected with the Application of the
Inductive Method of Reasoning to the Science of Medicine. Post 8vo. cloth, 5s.

III.

GOUT AND RHEUMATISM IN RELATION TO DISEASE
OF THE HEART. Post 8vo. cloth, 5s.

DR. T. HERBERT BARKER, M.D., F.R.S., & MR. ERNEST EDWARDS, B.A.

PHOTOGRAPHS OF EMINENT MEDICAL MEN, with brief
Analytical Notices of their Works. Nos. I. to VIII., price 3s. each.

DR. W. G. BARKER, M.D.LOND.

ON DISEASES OF THE RESPIRATORY PASSAGES AND
LUNGS, SPORADIC AND EPIDEMIC; their *Causes*, Pathology, Symptoms,
and Treatment. Crown 8vo. cloth, 6s.

DR. BARLOW.

A MANUAL OF THE PRACTICE OF MEDICINE. Second
Edition. Fcap. 8vo. cloth, 12s. 6d.

DR. BARNES.

THE PHYSIOLOGY AND TREATMENT OF PLACENTA
PRÆVIA; being the Lettsomian Lectures on Midwifery for 1857. Post 8vo. cloth, 6s.

DR. BASCOME.

A HISTORY OF EPIDEMIC PESTILENCES, FROM THE
EARLIEST AGES. 8vo. cloth, 8s.

DR. BASHAM.

ON DROPSY, AND ITS CONNECTION WITH DISEASES OF
THE KIDNEYS, HEART, LUNGS AND LIVER. With 16 Plates. Third
Edition. 8vo. cloth, 12s. 6d.

MR. H. F. BAXTER, M.R.C.S.L.

ON ORGANIC POLARITY; showing a Connexion to exist between
Organic Forces and Ordinary Polar Forces. Crown 8vo. cloth, 5s.

MR. BATEMAN.

MAGNACOPIA: A Practical Library of Profitable Knowledge, communicating the general Minutiæ of Chemical and Pharmaceutic Routine, together with the
generality of Secret Forms of Preparations. Third Edition. 18mo. 6s.

MR. LIONEL J. BEALE, M.R.C.S.

I.

THE LAWS OF HEALTH IN THEIR RELATIONS TO MIND
AND BODY. A Series of Letters from an Old Practitioner to a Patient. Post 8vo.
cloth, 7s. 6d.

II.

HEALTH AND DISEASE, IN CONNECTION WITH THE
GENERAL PRINCIPLES OF HYGIENE. Fcap. 8vo., 2s. 6d.

DR. BEALE, F.R.S.

I.

URINE, URINARY DEPOSITS, AND CALCULI: and on the Treatment of Urinary Diseases. Numerous Engravings. Second Edition, much Enlarged. Post 8vo. cloth, 8s. 6d.

II.

THE MICROSCOPE, IN ITS APPLICATION TO PRACTICAL MEDICINE. With a Coloured Plate, and 270 Woodcuts. Second Edition. 8vo. cloth, 14s.

III.

ILLUSTRATIONS OF THE SALTS OF URINE, URINARY DEPOSITS, and CALCULI. 37 Plates, containing upwards of 170 Figures copied from Nature, with descriptive Letterpress. 8vo. cloth, 9s. 6d.

MR. BEASLEY.

I.

THE BOOK OF PRESCRIPTIONS; containing 3000 Prescriptions. Collected from the Practice of the most eminent Physicians and Surgeons, English and Foreign. Third Edition. 18mo. cloth, 6s.

II.

THE DRUGGIST'S GENERAL RECEIPT-BOOK: comprising a copious Veterinary Formulary and Table of Veterinary Materia Medica; Patent and Proprietary Medicines, Druggists' Nostrums, &c.; Perfumery, Skin Cosmetics, Hair Cosmetics, and Teeth Cosmetics; Beverages, Dietetic Articles, and Condiments; Trade Chemicals, Miscellaneous Preparations and Compounds used in the Arts, &c.; with useful Memoranda and Tables. Sixth Edition. 18mo. cloth, 6s.

III.

THE POCKET FORMULARY AND SYNOPSIS OF THE BRITISH AND FOREIGN PHARMACOPŒIAS; comprising standard and approved Formulæ for the Preparations and Compounds employed in Medical Practice. Eighth Edition, corrected and enlarged. 18mo. cloth, 6s.

DR. HENRY BENNET.

I.

A PRACTICAL TREATISE ON INFLAMMATION AND OTHER DISEASES OF THE UTERUS. Fourth Edition, revised, with Additions. 8vo. cloth, 16s.

II.

A REVIEW OF THE PRESENT STATE OF UTERINE PATHOLOGY. 8vo. cloth, 4s.

III.

WINTER IN THE SOUTH OF EUROPE; OR, MENTONE, THE RIVIERA, CORSICA, SICILY, AND BIARRITZ, AS WINTER CLIMATES. Third Edition, with numerous Plates, Maps, and Wood Engravings. Post 8vo. cloth, 10s. 6d.

PROFESSOR BENTLEY, F.L.S.

A MANUAL OF BOTANY. With nearly 1,200 Engravings on Wood. Fcap. 8vo. cloth, 12s. 6d.

DR. BERNAYS.

NOTES FOR STUDENTS IN CHEMISTRY; being a Syllabus compiled from the Manuals of Miller, Fownes, Berzelius, Gerhardt, Gorup-Besanez, &c. Fourth Edition. Fscap. 8vo. cloth, 3s.

MR. HENRY HEATHER BIGG.

ORTHOPRAXY: the Mechanical Treatment of Deformities, Debilities, and Deficiencies of the Human Frame. With Engravings. Post 8vo. cloth, 10s.

DR. BILLING, F.R.S.

ON DISEASES OF THE LUNGS AND HEART. 8vo. cloth, 6s.

DR. S. B. BIRCH, M.D.

CONSTIPATED BOWELS: the Various Causes and the Rational Means of Cure. Second Edition. Post 8vo. cloth, 3s. 6d.

DR. GOLDING BIRD, F.R.S.
I.

URINARY DEPOSITS; THEIR DIAGNOSIS, PATHOLOGY, AND THERAPEUTICAL INDICATIONS. With Engravings. Fifth Edition. Edited by E. Lloyd Birkett, M.D. Post 8vo. cloth, 10s. 6d.

II.

ELEMENTS OF NATURAL PHILOSOPHY; being an Experimental Introduction to the Study of the Physical Sciences. With numerous Engravings. Fifth Edition. Edited by Charles Brooke, M.B. Cantab., F.R.S. Fcap. 8vo. cloth, 12s. 6d.

MR. BISHOP, F.R.S.
I.

ON DEFORMITIES OF THE HUMAN BODY, their Pathology and Treatment. With Engravings on Wood. 8vo. cloth, 10s.

II.

ON ARTICULATE SOUNDS, AND ON THE CAUSES AND CURE OF IMPEDIMENTS OF SPEECH. 8vo. cloth, 4s.

MR. P. HINCKES BIRD, F.R.C.S.

PRACTICAL TREATISE ON THE DISEASES OF CHILDREN AND INFANTS AT THE BREAST. Translated from the French of M. Bouchut, with Notes and Additions. 8vo. cloth. 20s.

MR. BLAINE.

OUTLINES OF THE VETERINARY ART; OR, A TREATISE ON THE ANATOMY, PHYSIOLOGY, AND DISEASES OF THE HORSE, NEAT CATTLE, AND SHEEP. Seventh Edition. By Charles Steel, M.R.C.V.S.L. With Plates. 8vo. cloth, 18s.

DR. BOURGUIGNON.

ON THE CATTLE PLAGUE; OR, CONTAGIOUS TYPHUS IN HORNED CATTLE: its History, Origin, Description, and Treatment. Post 8vo. 5s.

MR. JOHN E. BOWMAN, & MR. C. L. BLOXAM.
I.

PRACTICAL CHEMISTRY, including Analysis. With numerous Illustrations on Wood. Fifth Edition. Foolscap 8vo. cloth, 6s. 6d.

II.

MEDICAL CHEMISTRY; with Illustrations on Wood. Fourth Edition, carefully revised. Fcap. 8vo. cloth, 6s. 6d.

DR. JAMES BRIGHT.

ON DISEASES OF THE HEART, LUNGS, & AIR PASSAGES;
with a Review of the several Climates recommended in these Affections. Third Edition. Post 8vo. cloth, 9s.

DR. BRINTON, F.R.S.
I.
THE DISEASES OF THE STOMACH, with an Introduction on its
Anatomy and Physiology; being Lectures delivered at St. Thomas's Hospital. Second Edition. 8vo. cloth, 10s. 6d.
II.
THE SYMPTOMS, PATHOLOGY, AND TREATMENT OF
ULCER OF THE STOMACH. Post 8vo. cloth, 5s.

MR. BERNARD E. BRODHURST, F.R.C.S.
I.
CURVATURES OF THE SPINE: their Causes, Symptoms, Pathology,
and Treatment. Second Edition. Roy. 8vo. cloth, with Engravings, 7s. 6d.
II.
ON THE NATURE AND TREATMENT OF CLUBFOOT AND
ANALOGOUS DISTORTIONS involving the TIBIO-TARSAL ARTICULATION. With Engravings on Wood. 8vo. cloth, 4s. 6d.
III.
PRACTICAL OBSERVATIONS ON THE DISEASES OF THE
JOINTS INVOLVING ANCHYLOSIS, and on the TREATMENT for the RESTORATION of MOTION. Third Edition, much enlarged, 8vo. cloth, 4s. 6d.

MR. THOMAS BRYANT, F.R.C.S.
I.
ON THE DISEASES AND INJURIES OF THE JOINTS.
CLINICAL AND PATHOLOGICAL OBSERVATIONS. Post 8vo. cloth, 7s. 6d.
II.
THE SURGICAL DISEASES OF CHILDREN. The Lettsomian
Lectures, delivered March, 1863. Post 8vo. cloth, 5s.

DR. BRYCE.

ENGLAND AND FRANCE BEFORE SEBASTOPOL, looked at
from a Medical Point of View. 8vo. cloth, 6s.

DR. BUCKLE, M.D., L.R.C.P.LOND.

VITAL AND ECONOMICAL STATISTICS OF THE HOSPITALS,
INFIRMARIES, &c., OF ENGLAND AND WALES. Royal 8vo. 5s.

DR. BUDD, F.R.S.
I.
ON DISEASES OF THE LIVER.
Illustrated with Coloured Plates and Engravings on Wood. Third Edition. 8vo. cloth, 16s.
II.
ON THE ORGANIC DISEASES AND FUNCTIONAL DIS-
ORDERS OF THE STOMACH. 8vo. cloth, 9s.

DR. JOHN CHARLES BUCKNILL, F.R.S., & DR. DANIEL H. TUKE.

A MANUAL OF PSYCHOLOGICAL MEDICINE: containing
the History, Nosology, Description, Statistics, Diagnosis, Pathology, and Treatment of
Insanity. Second Edition. 8vo. cloth, 15s.

MR. CALLENDER, F.R.C.S.

FEMORAL RUPTURE: Anatomy of the Parts concerned. With Plates.
8vo. cloth, 4s.

DR. JOHN M. CAMPLIN, F.L.S.

ON DIABETES, AND ITS SUCCESSFUL TREATMENT.
Third Edition, by Dr. Glover. Fcap. 8vo. cloth, 3s. 6d.

MR. ROBERT B. CARTER, M.R.C.S.

I.
ON THE INFLUENCE OF EDUCATION AND TRAINING
IN PREVENTING DISEASES OF THE NERVOUS SYSTEM. Fcap. 8vo., 6s.

II.
THE PATHOLOGY AND TREATMENT OF HYSTERIA. Post
8vo. cloth, 4s. 6d.

DR. CARPENTER, F.R.S.

I.
PRINCIPLES OF HUMAN PHYSIOLOGY. With numerous Illus-
trations on Steel and Wood. Sixth Edition. Edited by Mr. HENRY POWER. 8vo.
cloth, 26s.

II.
A MANUAL OF PHYSIOLOGY. With 252 Illustrations on Steel
and Wood. Fourth Edition. Fcap. 8vo. cloth, 12s. 6d.

III.
THE MICROSCOPE AND ITS REVELATIONS. With nume-
rous Engravings on Steel and Wood. Third Edition. Fcap. 8vo. cloth, 12s. 6d.

DR. CHAMBERS.

I.
LECTURES, CHIEFLY CLINICAL. Fourth Edition. 8vo. cloth, 14s.

II.
DIGESTION AND ITS DERANGEMENTS. Post 8vo. cloth, 10s. 6d.

III.
SOME OF THE EFFECTS OF THE CLIMATE OF ITALY.
Crown 8vo. cloth, 4s. 6d.

DR. CHANCE, M.B.

VIRCHOW'S CELLULAR PATHOLOGY, AS BASED UPON
PHYSIOLOGICAL AND PATHOLOGICAL HISTOLOGY. With 144 Engravings on Wood. 8vo. cloth, 16s.

MR. H. T. CHAPMAN, F.R.C.S.

I.
THE TREATMENT OF OBSTINATE ULCERS AND CUTA-
NEOUS ERUPTIONS OF THE LEG WITHOUT CONFINEMENT. Third
Edition. Post 8vo. cloth, 3s. 6d.

II.
VARICOSE VEINS: their Nature, Consequences, and Treatment, Pallia-
tive and Curative. Second Edition. Post 8vo. cloth, 3s. 6d.

MR. PYE HENRY CHAVASSE, F.R.C.S.

I.

ADVICE TO A MOTHER ON THE MANAGEMENT OF
HER CHILDREN. Eighth Edition. Foolscap 8vo., 2s. 6d.

II.

ADVICE TO A WIFE ON THE MANAGEMENT OF HER
OWN HEALTH. With an Introductory Chapter, especially addressed to a Young Wife. Seventh Edition. Fcap. 8vo., 2s. 6d.

MR. LE GROS CLARK, F.R.C.S

OUTLINES OF SURGERY ; being an Epitome of the Lectures on the
Principles and the Practice of Surgery, delivered at St. Thomas's Hospital. Fcap. 8vo. cloth, 5s.

MR. JOHN CLAY, M.R.C.S.

KIWISCH ON DISEASES OF THE OVARIES: Translated, by
permission, from the last German Edition of his Clinical Lectures on the Special Pathology and Treatment of the Diseases of Women. With Notes, and an Appendix on the Operation of Ovariotomy. Royal 12mo. cloth, 16s.

DR. COCKLE, M.D.

ON INTRA-THORACIC CANCER. 8vo. 6s. 6d.

MR. COLLIS, M.B.DUB., F.R.C.S.I.

THE DIAGNOSIS AND TREATMENT OF CANCER AND
THE TUMOURS ANALOGOUS TO IT. With coloured Plates. 8vo. cloth, 14s.

DR. CONOLLY.

THE CONSTRUCTION AND GOVERNMENT OF LUNATIC
ASYLUMS AND HOSPITALS FOR THE INSANE. With Plans. Post 8vo. cloth, 6s.

MR. COOLEY.

COMPREHENSIVE SUPPLEMENT TO THE PHARMACOPŒIAS.

THE CYCLOPÆDIA OF PRACTICAL RECEIPTS, PRO-
CESSES, AND COLLATERAL INFORMATION IN THE ARTS, MANU-FACTURES, PROFESSIONS, AND TRADES, INCLUDING MEDICINE, PHARMACY, AND DOMESTIC ECONOMY; designed as a General Book of Reference for the Manufacturer, Tradesman, Amateur, and Heads of Families. Fourth and greatly enlarged Edition, 8vo. cloth, 28s.

MR. W. WHITE COOPER.

I.

ON WOUNDS AND INJURIES OF THE EYE. Illustrated by
17 Coloured Figures and 41 Woodcuts. 8vo. cloth, 12s.

II.

ON NEAR SIGHT, AGED SIGHT, IMPAIRED VISION,
AND THE MEANS OF ASSISTING SIGHT. With 31 Illustrations on Wood. Second Edition. Fcap. 8vo. cloth, 7s. 6d.

SIR ASTLEY COOPER, BART., F.R.S.
ON THE STRUCTURE AND DISEASES OF THE TESTIS.
With 24 Plates. Second Edition. Royal 4to., 20s.

MR. COOPER.
A DICTIONARY OF PRACTICAL SURGERY AND ENCYCLO-
PÆDIA OF SURGICAL SCIENCE. New Edition, brought down to the present time. By SAMUEL A. LANE, F.R.C.S., assisted by various eminent Surgeons. Vol. I., 8vo. cloth, £1. 5s.

MR. HOLMES COOTE, F.R.C.S.
A REPORT ON SOME IMPORTANT POINTS IN THE
TREATMENT OF SYPHILIS. 8vo. cloth, 5s.

DR. COTTON.
I.
ON CONSUMPTION : Its Nature, Symptoms, and Treatment. To
which Essay was awarded the Fothergillian Gold Medal of the Medical Society of London. Second Edition. 8vo. cloth, 8s.

II.
PHTHISIS AND THE STETHOSCOPE; OR, THE PHYSICAL
SIGNS OF CONSUMPTION. Third Edition. Foolscap 8vo. cloth, 3s.

MR. COULSON.
I.
ON DISEASES OF THE BLADDER AND PROSTATE GLAND.
New Edition, revised. *In Preparation.*
II.
ON LITHOTRITY AND LITHOTOMY; with Engravings on Wood.
8vo. cloth, 8s.

MR. WILLIAM CRAIG, L.F.P.S., GLASGOW.
ON THE INFLUENCE OF VARIATIONS OF ELECTRIC
TENSION AS THE REMOTE CAUSE OF EPIDEMIC AND OTHER DISEASES. 8vo. cloth, 10s.

MR. CURLING, F.R.S.
I.
OBSERVATIONS ON DISEASES OF THE RECTUM. Third
Edition. 8vo. cloth, 7s. 6d.
II.
A PRACTICAL TREATISE ON DISEASES OF THE TESTIS,
SPERMATIC CORD, AND SCROTUM. Third Edition, with Engravings. 8vo. cloth, 16s.

DR. DALRYMPLE, M.R.C.P., F.R.C.S.
THE CLIMATE OF EGYPT: METEOROLOGICAL AND MEDI-
CAL OBSERVATIONS, with Practical Hints for Invalid Travellers. Post 8vo. cloth, 4s.

MR. JOHN DALRYMPLE, F.R.S., F.R.C.S.
PATHOLOGY OF THE HUMAN EYE. Complete in Nine Fasciculi:
imperial 4to., 20s. each; half-bound morocco, gilt tops, 9l. 15s.

DR. HERBERT DAVIES.

ON THE PHYSICAL DIAGNOSIS OF DISEASES OF THE LUNGS AND HEART. Second Edition. Post 8vo. cloth, 8s.

DR. DAVEY.

I.

THE GANGLIONIC NERVOUS SYSTEM: its Structure, Functions, and Diseases. 8vo. cloth, 9s.

II.

ON THE NATURE AND PROXIMATE CAUSE OF IN-SANITY. Post 8vo. cloth, 3s.

DR. HENRY DAY, M.D., M.R.C.P.

CLINICAL HISTORIES; with Comments. 8vo. cloth, 7s. 6d.

MR. DIXON.

A GUIDE TO THE PRACTICAL STUDY OF DISEASES OF THE EYE. Third Edition. Post 8vo. cloth, 9s.

DR. DOBELL.

I.

DEMONSTRATIONS OF DISEASES IN THE CHEST, AND THEIR PHYSICAL DIAGNOSIS. With Coloured Plates. 8vo. cloth, 12s. 6d.

II.

LECTURES ON THE GERMS AND VESTIGES OF DISEASE, and on the Prevention of the Invasion and Fatality of Disease by Periodical Examinations. 8vo. cloth, 6s. 6d.

III.

A MANUAL OF DIET AND REGIMEN FOR PHYSICIAN AND PATIENT. Third Edition (for the year 1865). Crown 8vo. cloth, 1s. 6d.

IV.

ON TUBERCULOSIS: ITS NATURE, CAUSE, AND TREAT-MENT; with Notes on Pancreatic Juice. Second Edition. Crown 8vo. cloth, 3s. 6d.

V.

ON WINTER COUGH (CATARRH, BRONCHITIS, EMPHY-SEMA, ASTHMA); with an Appendix on some Principles of Diet in Disease— Lectures delivered at the Royal Infirmary for Diseases of the Chest. Post 8vo. cloth, 5s. 6d.

DR. TOOGOOD DOWNING.

NEURALGIA: its various Forms, Pathology, and Treatment. THE JACKSONIAN PRIZE ESSAY FOR 1850. 8vo. cloth, 10s. 6d.

DR. DRUITT, F.R.C.S.

THE SURGEON'S VADE-MECUM; with numerous Engravings on Wood. Ninth Edition. Foolscap 8vo. cloth, 12s. 6d.

MR. DUNN, F.R.C.S.

AN ESSAY ON PHYSIOLOGICAL PSYCHOLOGY. 8vo. cloth, 4s.

SIR JAMES EYRE, M.D.

I.

THE STOMACH AND ITS DIFFICULTIES. Fifth Edition. Fcap. 8vo. cloth, 2s. 6d.

II.

PRACTICAL REMARKS ON SOME EXHAUSTING DIS-EASES. Second Edition. Post 8vo. cloth, 4s. 6d.

DR. FAYRER, M.D., F.R.C.S.

CLINICAL SURGERY IN INDIA. With Engravings. 8vo. cloth, 16s.

DR. FENWICK.

ON SCROFULA AND CONSUMPTION. Clergyman's Sore Throat, Catarrh, Croup, Bronchitis, Asthma. Fcap. 8vo., 2s. 6d.

SIR WILLIAM FERGUSSON, BART., F.R.S.

A SYSTEM OF PRACTICAL SURGERY; with numerous Illustrations on Wood. Fourth Edition. Fcap. 8vo. cloth, 12s. 6d.

SIR JOHN FIFE, F.R.C.S. AND MR. URQUHART.

MANUAL OF THE TURKISH BATH. Heat a Mode of Cure and a Source of Strength for Men and Animals. With Engravings. Post 8vo. cloth, 5s.

MR. FLOWER, F.R.S., F.R.C.S.

DIAGRAMS OF THE NERVES OF THE HUMAN BODY, exhibiting their Origin, Divisions, and Connexions, with their Distribution to the various Regions of the Cutaneous Surface, and to all the Muscles. Folio, containing Six Plates, 14s.

MR. FOWNES, PH.D., F.R.S.

I.

A MANUAL OF CHEMISTRY; with 187 Illustrations on Wood. Ninth Edition. Fcap. 8vo. cloth, 12s. 6d.
Edited by H. Bence Jones, M.D., F.R.S., and A. W. Hofmann, Ph.D., F.R.S.

II.

CHEMISTRY, AS EXEMPLIFYING THE WISDOM AND BENEFICENCE OF GOD. Second Edition. Fcap. 8vo. cloth, 4s. 6d.

III.

INTRODUCTION TO QUALITATIVE ANALYSIS. Post 8vo. cloth, 2s.

DR. D. J. T. FRANCIS.

CHANGE OF CLIMATE; considered as a Remedy in Dyspeptic, Pulmonary, and other Chronic Affections; with an Account of the most Eligible Places of Residence for Invalids, at different Seasons of the Year. Post 8vo. cloth, 8s. 6d.

DR. W. FRAZER.

ELEMENTS OF MATERIA MEDICA; containing the Chemistry and Natural History of Drugs—their Effects, Doses, and Adulterations. Second Edition. 8vo. cloth, 10s. 6d.

C. REMIGIUS FRESENIUS.

A SYSTEM OF INSTRUCTION IN CHEMICAL ANALYSIS, Edited by Lloyd Bullock, F.C.S.
Qualitative. Sixth Edition, with Coloured Plate illustrating Spectrum Analysis. 8vo. cloth, 10s. 6d.——Quantitative. Fourth Edition. 8vo. cloth, 18s.

DR. FULLER.
I.

ON DISEASES OF THE CHEST, including Diseases of the Heart and Great Vessels. With Engravings. 8vo. cloth, 12s. 6d.

II.

ON DISEASES OF THE HEART AND GREAT VESSELS. 8vo. cloth, 7s. 6d.

III.

ON RHEUMATISM, RHEUMATIC GOUT, AND SCIATICA: their Pathology, Symptoms, and Treatment. Third Edition. 8vo. cloth, 12s. 6d.

DR. GAIRDNER.

ON GOUT; its History, its Causes, and its Cure. Fourth Edition. Post 8vo. cloth, 8s. 6d.

MR. GALLOWAY.
I.

THE FIRST STEP IN CHEMISTRY. Third Edition. Fcap. 8vo. cloth, 5s.

II.

THE SECOND STEP IN CHEMISTRY; or, the Student's Guide to the Higher Branches of the Science. With Engravings. 8vo. cloth, 10s.

III.

A MANUAL OF QUALITATIVE ANALYSIS. Fourth Edition. Post 8vo. cloth, 6s. 6d.

IV.

CHEMICAL TABLES. On Five Large Sheets, for School and Lecture Rooms. Second Edition. 4s. 6d.

MR. J. SAMPSON GAMGEE.

HISTORY OF A SUCCESSFUL CASE OF AMPUTATION AT THE HIP-JOINT (the limb 48-in. in circumference, 99 pounds weight). With 4 Photographs. 4to cloth, 10s. 6d.

MR. F. J. GANT, F.R.C.S.
I.

THE PRINCIPLES OF SURGERY: Clinical, Medical, and Operative. With Engravings. 8vo. cloth, 18s.

II.

THE IRRITABLE BLADDER: its Causes and Curative Treatment. Post 8vo. cloth, 4s. 6d.

DR. GIBB, M.R.C.P.

ON DISEASES OF THE THROAT AND WINDPIPE, as reflected by the Laryngoscope. Second Edition. With 116 Engravings. Post 8vo. cloth, 10s. 6d.

MRS. GODFREY.

ON THE NATURE, PREVENTION, TREATMENT, AND CURE OF SPINAL CURVATURES and DEFORMITIES of the CHEST and LIMBS, without ARTIFICIAL SUPPORTS or any MECHANICAL APPLIANCES. Third Edition, Revised and Enlarged. 8vo. cloth, 5s.

DR. GORDON, M.D., C.B.
I.

ARMY HYGIENE. 8vo. cloth, 20s.

II.

CHINA, FROM A MEDICAL POINT OF VIEW, IN 1860 AND 1861; With a Chapter on Nagasaki as a Sanatarium. 8vo. cloth, 10s. 6d.

DR. GRANVILLE, F.R.S.

I.

THE MINERAL SPRINGS OF VICHY : their Efficacy in the
Treatment of Gout, Indigestion, Gravel, &c. 8vo. cloth, 3s,

II.

ON SUDDEN DEATH.   Post 8vo., 2s. 6d.

DR. GRAVES, M.D., F.R.S.

STUDIES IN PHYSIOLOGY AND MEDICINE. Edited by
Dr. Stokes. With Portrait and Memoir. 8vo. cloth, 14s.

DR. S. C. GRIFFITH, M.D.

ON DERMATOLOGY AND THE TREATMENT OF SKIN
DISEASES BY MEANS OF HERBS, IN PLACE OF ARSENIC AND
MERCURY. Fcap. 8vo. cloth, 3s.

MR. GRIFFITHS.

CHEMISTRY OF THE FOUR SEASONS — Spring, Summer,
Autumn, Winter. Illustrated with Engravings on Wood. Second Edition. Foolscap
8vo. cloth, 7s. 6d.

DR. GULLY.

THE SIMPLE TREATMENT OF DISEASE; deduced from the
Methods of Expectancy and Revulsion. 18mo. cloth, 4s.

DR. GUY AND DR. JOHN HARLEY.

HOOPER'S PHYSICIAN'S VADE-MECUM; OR, MANUAL OF
THE PRINCIPLES AND PRACTICE OF PHYSIC. Seventh Edition, considerably enlarged, and rewritten. Foolscap 8vo. cloth, 12s. 6d.

GUY'S HOSPITAL REPORTS. Third Series. Vols. I. to XII., 8vo.,
7s. 6d. each.

DR. HABERSHON, F.R.C.P.

I.

ON DISEASES OF THE ABDOMEN, comprising those of the
Stomach and other Parts of the Alimentary Canal, Œsophagus, Stomach, Cæcum,
Intestines, and Peritoneum. Second Edition, with Plates. 8vo. cloth, 14s.

II.

ON THE INJURIOUS EFFECTS OF MERCURY IN THE
TREATMENT OF DISEASE. Post 8vo. cloth, 3s. 6d.

DR. C. RADCLYFFE HALL.

TORQUAY IN ITS MEDICAL ASPECT AS A RESORT FOR
PULMONARY INVALIDS. Post 8vo. cloth, 5s.

DR. MARSHALL HALL, F.R.S.

I.

PRONE AND POSTURAL RESPIRATION IN DROWNING
AND OTHER FORMS OF APNŒA OR SUSPENDED RESPIRATION.
Post 8vo. cloth. 5s.

II.

PRACTICAL OBSERVATIONS AND SUGGESTIONS IN MEDI-
CINE. Second Series. Post 8vo. cloth, 8s. 6d.

MR. HARDWICH.

A MANUAL OF PHOTOGRAPHIC CHEMISTRY. With Engravings. Seventh Edition. Foolscap 8vo. cloth, 7s. 6d.

DR. J. BOWER HARRISON, M.D., M.R.C.P.

I.

LETTERS TO A YOUNG PRACTITIONER ON THE DIS-EASES OF CHILDREN. Foolscap 8vo. cloth, 5s.

II.

ON THE CONTAMINATION OF WATER BY THE POISON OF LEAD, and its Effects on the Human Body. Foolscap 8vo. cloth, 3s. 6d.

DR. HARTWIG.

I.

ON SEA BATHING AND SEA AIR. Second Edition. Fcap. 8vo., 2s. 6d.

II.

ON THE PHYSICAL EDUCATION OF CHILDREN. Fcap. 8vo., 2s. 6d.

DR. A. H. HASSALL.

I.

THE URINE, IN HEALTH AND DISEASE; being an Explanation of the Composition of the Urine, and of the Pathology and Treatment of Urinary and Renal Disorders. Second Edition. With 79 Engravings (23 Coloured). Post 8vo. cloth, 12s. 6d.

II.

THE MICROSCOPIC ANATOMY OF THE HUMAN BODY, IN HEALTH AND DISEASE. Illustrated with Several Hundred Drawings in Colour. Two vols. 8vo. cloth, £1. 10s.

MR. ALFRED HAVILAND, M.R.C.S.

CLIMATE, WEATHER, AND DISEASE; being a Sketch of the Opinions of the most celebrated Ancient and Modern Writers with regard to the Influence of Climate and Weather in producing Disease. With Four coloured Engravings. 8vo. cloth, 7s.

DR. HEADLAND.

ON THE ACTION OF MEDICINES IN THE SYSTEM. Third Edition. 8vo. cloth, 12s. 6d.

DR. HEALE.

I.

A TREATISE ON THE PHYSIOLOGICAL ANATOMY OF THE LUNGS. With Engravings. 8vo. cloth, 8s.

II.

A TREATISE ON VITAL CAUSES. 8vo. cloth, 9s.

MR. CHRISTOPHER HEATH, F.R.C.S.

I.

PRACTICAL ANATOMY: a Manual of Dissections. With numerous Engravings. Fcap. 8vo. cloth, 10s. 6d.

II.

A MANUAL OF MINOR SURGERY AND BANDAGING, FOR THE USE OF HOUSE-SURGEONS, DRESSERS, AND JUNIOR PRACTITIONERS. With Illustrations. Third Edition. Fcap. 8vo. cloth, 5s.

MR. HIGGINBOTTOM, F.R.S., F.R.C.S.E.

A PRACTICAL ESSAY ON THE USE OF THE NITRATE OF SILVER IN THE TREATMENT OF INFLAMMATION, WOUNDS, AND ULCERS. Third Edition, 8vo. cloth, 6s.

DR. HINDS.

THE HARMONIES OF PHYSICAL SCIENCE IN RELATION TO THE HIGHER SENTIMENTS; with Observations on Medical Studies, and on the Moral and Scientific Relations of Medical Life. Post 8vo. cloth, 4s.

MR. J. A. HINGESTON, M.R.C.S.

TOPICS OF THE DAY, MEDICAL, SOCIAL, AND SCIENTIFIC. Crown 8vo. cloth, 7s. 6d.

DR. HODGES.

THE NATURE, PATHOLOGY, AND TREATMENT OF PUERPERAL CONVULSIONS. Crown 8vo. cloth, 3s.

DR. DECIMUS HODGSON.

THE PROSTATE GLAND, AND ITS ENLARGEMENT IN OLD AGE. With 12 Plates. Royal 8vo. cloth, 6s.

MR. JABEZ HOGG.

A MANUAL OF OPHTHALMOSCOPIC SURGERY; being a Practical Treatise on the Use of the Ophthalmoscope in Diseases of the Eye. Third Edition. With Coloured Plates. 8vo. cloth, 10s. 6d.

MR. LUTHER HOLDEN, F.R.C.S.

I.

HUMAN OSTEOLOGY: with Plates, showing the Attachments of the Muscles. Third Edition. 8vo. cloth, 16s.

II.

A MANUAL OF THE DISSECTION OF THE HUMAN BODY. With Engravings on Wood. Second Edition. 8vo. cloth, 16s.

MR BARNARD HOLT, F.R.C.S.

ON THE IMMEDIATE TREATMENT OF STRICTURE OF THE URETHRA. Second Edition, Enlarged. 8vo. cloth, 3s.

DR. W. CHARLES HOOD.

SUGGESTIONS FOR THE FUTURE PROVISION OF CRIMINAL LUNATICS. 8vo. cloth, 5s. 6d.

DR. P. HOOD.

THE SUCCESSFUL TREATMENT OF SCARLET FEVER; also, OBSERVATIONS ON THE PATHOLOGY AND TREATMENT OF CROWING INSPIRATIONS OF INFANTS. Post 8vo. cloth, 5s.

MR. JOHN HORSLEY.

A CATECHISM OF CHEMICAL PHILOSOPHY; being a Familiar Exposition of the Principles of Chemistry and Physics. With Engravings on Wood. Designed for the Use of Schools and Private Teachers. Post 8vo. cloth, 6s. 6d.

MR. LUKE HOWARD, F.R.S.

ESSAY ON THE MODIFICATIONS OF CLOUDS. Third Edition, by W. D. and E. HOWARD. With 6 Lithographic Plates, from Pictures by Kenyon. 4to. cloth, 10s. 6d.

DR. HAMILTON HOWE, M.D.

A THEORETICAL INQUIRY INTO THE PHYSICAL CAUSE OF EPIDEMIC DISEASES. Accompanied with Tables. 8vo. cloth, 7s.

DR. HUFELAND.

THE ART OF PROLONGING LIFE. Second Edition. Edited by ERASMUS WILSON, F.R.S. Foolscap 8vo., 2s. 6d.

MR. W. CURTIS HUGMAN, F.R.C.S.

ON HIP-JOINT DISEASE; with reference especially to Treatment by Mechanical Means for the Relief of Contraction and Deformity of the Affected Limb. With Plates. Re-issue, enlarged. 8vo. cloth, 3s. 6d.

MR. HULKE, F.R.C.S.

A PRACTICAL TREATISE ON THE USE OF THE OPHTHALMOSCOPE. Being the Jacksonian Prize Essay for 1859. Royal 8vo. cloth, 8s.

DR. HENRY HUNT.

ON HEARTBURN AND INDIGESTION. 8vo. cloth, 5s.

PROFESSOR HUXLEY, F.R.S.

LECTURES ON THE ELEMENTS OF COMPARATIVE ANATOMY.—ON CLASSIFICATON AND THE SKULL. With 111 Illustrations. 8vo. cloth, 10s. 6d.

MR. JONATHAN HUTCHINSON, F.R.C.S.

A CLINICAL MEMOIR ON CERTAIN DISEASES OF THE
EYE AND EAR, CONSEQUENT ON INHERITED SYPHILIS; with an
appended Chapter of Commentaries on the Transmission of Syphilis from Parent to
Offspring, and its more remote Consequences.   With Plates and Woodcuts, 8vo. cloth, 9s.

DR. INMAN, M.R.C.P.

I.

ON MYALGIA: ITS NATURE, CAUSES, AND TREATMENT;
being a Treatise on Painful and other Affections of the Muscular System.   Second
Edition. 8vo. cloth, 9s.

II.

FOUNDATION FOR A NEW THEORY AND PRACTICE
OF MEDICINE.   Second Edition.   Crown 8vo. cloth, 10s.

DR. JAGO, M.D.OXON., A.B.CANTAB.

ENTOPTICS, WITH ITS USES IN PHYSIOLOGY AND
MEDICINE.   With 54 Engravings.   Crown 8vo. cloth, 5s.

MR. J. H. JAMES, F.R.C.S.

PRACTICAL OBSERVATIONS ON THE OPERATIONS FOR
STRANGULATED HERNIA.   8vo. cloth, 5s.

DR. PROSSER JAMES, M.D.

SORE-THROAT: ITS NATURE, VARIETIES, AND TREAT-
MENT; including the Use of the LARYNGOSCOPE as an Aid to Diagnosis.   Second
Edition, with numerous Engravings.   Post 8vo. cloth, 5s.

DR. HANDFIELD JONES, M.B., F.R.C.P.

CLINICAL OBSERVATIONS ON FUNCTIONAL NERVOUS
DISORDERS.   Post 8vo. cloth, 10s. 6d.

DR. HANDFIELD JONES, F.R.S., & DR. EDWARD H. SIEVEKING.

A MANUAL OF PATHOLOGICAL ANATOMY.   Illustrated with
numerous Engravings on Wood.   Foolscap 8vo. cloth, 12s. 6d.

DR. JAMES JONES, M.D., M.R.C.P.

ON THE USE OF PERCHLORIDE OF IRON AND OTHER
CHALYBEATE SALTS IN THE TREATMENT OF CONSUMPTION.   Crown
8vo. cloth, 3s. 6d.

MR. WHARTON JONES, F.R.S.

I.

A MANUAL OF THE PRINCIPLES AND PRACTICE OF
OPHTHALMIC MEDICINE AND SURGERY; with Nine Coloured Plates and 173 Wood Engravings. Third Edition, thoroughly revised. Foolscap 8vo. cloth, 12s. 6d.

II.

THE WISDOM AND BENEFICENCE OF THE ALMIGHTY,
AS DISPLAYED IN THE SENSE OF VISION. Actonian Prize Essay. With Illustrations on Steel and Wood. Foolscap 8vo. cloth, 4s. 6d.

III.

DEFECTS OF SIGHT AND HEARING: their Nature, Causes, Prevention, and General Management. Second Edition, with Engravings. Fcap. 8vo. 2s. 6d.

IV.

A CATECHISM OF THE MEDICINE AND SURGERY OF
THE EYE AND EAR. For the Clinical Use of Hospital Students. Fcap. 8vo. 2s. 6d.

V.

A CATECHISM OF THE PHYSIOLOGY AND PHILOSOPHY
OF BODY, SENSE, AND MIND. For Use in Schools and Colleges. Fcap. 8vo., 2s. 6d.

MR. FURNEAUX JORDAN, M.R.C.S.

AN INTRODUCTION TO CLINICAL SURGERY; WITH A
Method of Investigating and Reporting Surgical Cases. Fcap. 8vo. cloth, 5s.

MR. JUDD.

A PRACTICAL TREATISE ON URETHRITIS AND SYPHI-
LIS: including Observations on the Power of the Menstruous Fluid, and of the Discharge from Leucorrhœa and Sores to produce Urethritis: with a variety of Examples, Experiments, Remedies, and Cures. 8vo. cloth, £1. 5s.

DR. LAENNEC.

A MANUAL OF AUSCULTATION AND PERCUSSION. Translated and Edited by J. B. Sharpe, M.R.C.S. 3s.

DR. LANE, M.A.

HYDROPATHY; OR, HYGIENIC MEDICINE. An Explanatory
Essay. Second Edition. Post 8vo. cloth, 5s.

MR. LAWRENCE, F.R.S.

I.

LECTURES ON SURGERY. 8vo. cloth, 16s.

II.

A TREATISE ON RUPTURES. The Fifth Edition, considerably
enlarged. 8vo. cloth, 16s.

DR. LEARED, M.R.C.P.

IMPERFECT DIGESTION: ITS CAUSES AND TREATMENT.
Fourth Edition. Foolscap 8vo. cloth, 4s.

DR. EDWIN LEE.

I.

THE EFFECT OF CLIMATE ON TUBERCULOUS DISEASE,
with Notices of the chief Foreign Places of Winter Resort. Small 8vo. cloth, 4s. 6d.

II.

THE WATERING PLACES OF ENGLAND, CONSIDERED
with Reference to their Medical Topography. Fourth Edition. Fcap. 8vo. cloth, 7s. 6d.

III.

THE PRINCIPAL BATHS OF FRANCE. Fourth Edition.
Fcap. 8vo. cloth, 3s. 6d.

IV.

THE BATHS OF GERMANY. Fourth Edition. Post 8vo. cloth, 7s.

V.

THE BATHS OF SWITZERLAND. 12mo. cloth, 3s. 6d.

VI.

HOMŒOPATHY AND HYDROPATHY IMPARTIALLY AP-
PRECIATED. With Notes illustrative of the Influence of the Mind over the Body.
Fourth Edition. Post 8vo. cloth, 3s. 6d.

MR. HENRY LEE, F.R.C.S.

I.

ON SYPHILIS. Second Edition. With Coloured Plates. 8vo. cloth, 10s.

II.

ON DISEASES OF THE VEINS, HÆMORRHOIDAL TUMOURS,
AND OTHER AFFECTIONS OF THE RECTUM. Second Edition. 8vo. cloth, 8s.

DR. ROBERT LEE, F.R.S.

I.

CONSULTATIONS IN MIDWIFERY. Foolscap 8vo. cloth, 4s. 6d.

II.

A TREATISE ON THE SPECULUM; with Three Hundred Cases.
8vo. cloth, 4s. 6d.

III.

CLINICAL REPORTS OF OVARIAN AND UTERINE DIS-
EASES, with Commentaries. Foolscap 8vo. cloth, 6s. 6d.

IV.

CLINICAL MIDWIFERY: comprising the Histories of 545 Cases of
Difficult, Preternatural, and Complicated Labour, with Commentaries. Second Edition.
Foolscap 8vo. cloth, 5s.

DR. LEISHMAN, M.D., F.F.P.S.

THE MECHANISM OF PARTURITION: An Essay, Historical and
Critical. With Engravings. 8vo. cloth, 5s.

MR. LISTON, F.R.S.

PRACTICAL SURGERY. Fourth Edition. 8vo. cloth, 22s.

MR. H. W. LOBB, L.S.A., M.R.C.S.E.

ON SOME OF THE MORE OBSCURE FORMS OF NERVOUS
AFFECTIONS, THEIR PATHOLOGY AND TREATMENT. Re-issue,
with the Chapter on Galvanism entirely Re-written. With Engravings. 8vo. cloth, 8s.

DR. LOGAN, M.D., M.R.C.P.LOND.

ON OBSTINATE DISEASES OF THE SKIN. Fcap. 8vo. cloth, 2s. 6d.

LONDON HOSPITAL.

CLINICAL LECTURES AND REPORTS BY THE MEDICAL
AND SURGICAL STAFF. With Illustrations. Vols. I. to III. 8vo. cloth, 7s. 6d.

LONDON MEDICAL SOCIETY OF OBSERVATION.

WHAT TO OBSERVE AT THE BED-SIDE, AND AFTER
DEATH. Published by Authority. Second Edition. Foolscap 8vo. cloth, 4s. 6d.

MR. M'CLELLAND, F.L.S., F.G.S.

THE MEDICAL TOPOGRAPHY, OR CLIMATE AND SOILS,
OF BENGAL AND THE N. W. PROVINCES. Post 8vo. cloth, 4s. 6d.

DR. MACLACHLAN, M.D., F.R.C.P.L.

THE DISEASES AND INFIRMITIES OF ADVANCED LIFE.
8vo. cloth, 16s.

DR. A. C. MACLEOD, M.R.C.P.LOND.

ACHOLIC DISEASES ; comprising Jaundice, Diarrhœa, Dysentery,
and Cholera. Post 8vo. cloth, 5s. 6d.

DR. GEORGE H. B. MACLEOD, F.R.C.S.E.

I.

OUTLINES OF SURGICAL DIAGNOSIS. 8vo. cloth, 12s. 6d.

II.

NOTES ON THE SURGERY OF THE CRIMEAN WAR; with
REMARKS on GUN-SHOT WOUNDS. 8vo. cloth, 10s. 6d.

MR. JOSEPH MACLISE, F.R.C.S.

I.

SURGICAL ANATOMY. A Series of Dissections, illustrating the Principal Regions of the Human Body.
The Second Edition, imperial folio, cloth, £3. 12s.; half-morocco, £4. 4s.

II.

ON DISLOCATIONS AND FRACTURES. This Work is Uniform
with the Author's "Surgical Anatomy;" each Fasciculus contains Four beautifully executed Lithographic Drawings. Imperial folio, cloth, £2. 10s.; half-morocco, £2. 17s.

MR. MACNAMARA.

ON DISEASES OF THE EYE; referring principally to those Affections
requiring the aid of the Ophthalmoscope for their Diagnosis. With coloured plates. 8vo. cloth, 10s. 6d.

DR. McNICOLL, M.R.C.P.

A HAND-BOOK FOR SOUTHPORT, MEDICAL & GENERAL;
with Copious Notices of the Natural History of the District. Second Edition. Post 8vo. cloth, 3s. 6d.

DR. MARCET, F.R.S.

I.

ON THE COMPOSITION OF FOOD, AND HOW IT IS
ADULTERATED; with Practical Directions for its Analysis. 8vo. cloth, 6s. 6d.

II.

ON CHRONIC ALCOHOLIC INTOXICATION; with an INQUIRY
INTO THE INFLUENCE OF THE ABUSE OF ALCOHOL AS A PREDISPOSING CAUSE OF DISEASE. Second Edition, much enlarged. Foolscap 8vo. cloth, 4s. 6d.

DR. J. MACPHERSON, M.D

CHOLERA IN ITS HOME; with a Sketch of the Pathology and Treatment of the Disease. Crown 8vo. cloth, 5s.

DR. MARKHAM.

I.

DISEASES OF THE HEART: THEIR PATHOLOGY, DIAGNOSIS, AND TREATMENT. Second Edition. Post 8vo. cloth, 6s.

II.

SKODA ON AUSCULTATION AND PERCUSSION. Post 8vo. cloth, 6s.

III.

BLEEDING AND CHANGE IN TYPE OF DISEASES. Gulstonian Lectures for 1864. Crown 8vo. 2s. 6d.

SIR RANALD MARTIN, K.C.B., F.R.S.

INFLUENCE OF TROPICAL CLIMATES IN PRODUCING THE ACUTE ENDEMIC DISEASES OF EUROPEANS; including Practical Observations on their Chronic Sequelæ under the Influences of the Climate of Europe. Second Edition, much enlarged. 8vo. cloth, 20s.

DR. MASSY.

ON THE EXAMINATION OF RECRUITS; intended for the Use of Young Medical Officers on Entering the Army. 8vo. cloth, 5s.

MR. C. F. MAUNDER, F.R.C.S.

OPERATIVE SURGERY. With 158 Engravings. Post 8vo. 6s.

DR. MAYNE.

I.

AN EXPOSITORY LEXICON OF THE TERMS, ANCIENT AND MODERN, IN MEDICAL AND GENERAL SCIENCE. 8vo. cloth, £2. 10s.

II.

A MEDICAL VOCABULARY; or, an Explanation of all Names, Synonymes, Terms, and Phrases used in Medicine and the relative branches of Medical Science. Second Edition. Fcap. 8vo. cloth, 8s. 6d.

DR. MERYON, M.D., F.R.C.P.

PATHOLOGICAL AND PRACTICAL RESEARCHES ON THE VARIOUS FORMS OF PARALYSIS. 8vo. cloth, 6s.

DR. MILLINGEN.

ON THE TREATMENT AND MANAGEMENT OF THE INSANE; with Considerations on Public and Private Lunatic Asylums. 18mo. cloth, 4s. 6d.

DR. W. J. MOORE, M.D.

I.

HEALTH IN THE TROPICS; or, Sanitary Art applied to Europeans in India. 8vo. cloth, 9s.

II.

A MANUAL OF THE DISEASES OF INDIA. Fcap. 8vo. cloth, 5s.

PROFESSOR MULDER, UTRECHT.

THE CHEMISTRY OF WINE. Edited by H. BENCE JONES, M.D., F.R.S. Fcap. 8vo. cloth, 6s.

DR. W. MURRAY, M.D., M.R.C.P.

EMOTIONAL DISORDERS OF THE SYMPATHETIC SYSTEM OF NERVES. Crown 8vo. cloth, 3s. 6d.

DR. MUSHET, M.B., M.R.C.P.

ON APOPLEXY, AND ALLIED AFFECTIONS OF THE BRAIN. 8vo. cloth, 7s.

MR. NAYLER, F.R.C.S.

ON THE DISEASES OF THE SKIN. With Plates. 8vo. cloth, 10s. 6d.

DR. BIRKBECK NEVINS.

THE PRESCRIBER'S ANALYSIS OF THE BRITISH PHARMACOPEIA. Third Edition, enlarged to 295 pp. 32mo. cloth, 3s. 6d.

DR. THOS. NICHOLSON, M.D.

ON YELLOW FEVER; comprising the History of that Disease as it appeared in the Island of Antigua. Fcap. 8vo. cloth, 2s. 6d.

DR. NOAD, PH.D., F.R.S.

THE INDUCTION COIL, being a Popular Explanation of the Electrical Principles on which it is constructed. Second Edition. With Engravings. Fcap. 8vo. cloth, 3s.

DR. NOBLE.

THE HUMAN MIND IN ITS RELATIONS WITH THE BRAIN AND NERVOUS SYSTEM. Post 8vo. cloth, 4s. 6d.

MR. NUNNELEY, F.R.C.S.E.

I.

ON THE ORGANS OF VISION: THEIR ANATOMY AND PHYSIOLOGY. With Plates, 8vo. cloth, 15s.

II.

A TREATISE ON THE NATURE, CAUSES, AND TREATMENT OF ERYSIPELAS. 8vo. cloth, 10s. 6d.

MR. LANGSTON PARKER.

THE MODERN TREATMENT OF SYPHILITIC DISEASES, both Primary and Secondary; comprising the Treatment of Constitutional and Confirmed Syphilis, by a safe and successful Method. Fourth Edition, 8vo. cloth, 10s.

DR. PARKES, F.R.S., F.R.C.P.

I.

A MANUAL OF PRACTICAL HYGIENE; intended especially for the Medical Officers of the Army. With Plates and Woodcuts. 2nd Edition, 8vo. cloth, 16s.

II.

THE URINE: ITS COMPOSITION IN HEALTH AND DISEASE, AND UNDER THE ACTION OF REMEDIES. 8vo. cloth, 12s.

DR. PARKIN, M.D., F.R.C.S.

I.

THE ANTIDOTAL TREATMENT AND PREVENTION OF
THE EPIDEMIC CHOLERA. Third Edition. 8vo. cloth, 7s. 6d.

II.

THE CAUSATION AND PREVENTION OF DISEASE; with
the Laws regulating the Extrication of Malaria from the Surface, and its Diffusion in the
surrounding Air. 8vo. cloth, 5s.

MR. JAMES PART, F.R.C.S.

THE MEDICAL AND SURGICAL POCKET CASE BOOK,
for the Registration of important Cases in Private Practice, and to assist the Student of
Hospital Practice. Second Edition. 2s. 6d.

DR. PAVY, M.D., F.R.S., F.R.C.P.

DIABETES : RESEARCHES ON ITS NATURE AND TREAT-
MENT. 8vo. cloth, 8s. 6d.

DR. PEACOCK, M.D., F.R.C.P.

ON SOME OF THE CAUSES AND EFFECTS OF VALVULAR
DISEASE OF THE HEART. With Engravings. 8vo. cloth, 5s.

DR. W. H. PEARSE, M.D.EDIN.

NOTES ON HEALTH IN CALCUTTA AND BRITISH
EMIGRANT SHIPS, including Ventilation, Diet, and Disease. Fcap. 8vo. 2s.

DR. PEET, M.D., F.R.C.P.

THE PRINCIPLES AND PRACTICE OF MEDICINE;
Designed chiefly for Students of Indian Medical Colleges. 8vo. cloth, 16s.

DR. PEREIRA, F.R.S.

SELECTA E PRÆSCRIPTIS. Fourteenth Edition. 24mo. cloth, 5s.

DR. PICKFORD.

HYGIENE; or, Health as Depending upon the Conditions of the Atmo-
sphere, Food and Drinks, Motion and Rest, Sleep and Wakefulness, Secretions, Excre-
tions, and Retentions, Mental Emotions, Clothing, Bathing, &c. Vol. I. 8vo. cloth, 9s.

MR. PIRRIE, F.R.S.E.

THE PRINCIPLES AND PRACTICE OF SURGERY. With
numerous Engravings on Wood. Second Edition. 8vo. cloth, 24s.

PHARMACOPŒIA COLLEGII REGALIS MEDICORUM LON-
DINENSIS. 8vo. cloth, 9s.; or 24mo. 5s.

PROFESSORS PLATTNER & MUSPRATT·

THE USE OF THE BLOWPIPE IN THE EXAMINATION OF
MINERALS, ORES, AND OTHER METALLIC COMBINATIONS. Illustrated
by numerous Engravings on Wood. Third Edition. 8vo. cloth, 10s. 6d.

DR. HENRY F. A. PRATT, M.D., M.R.C.P.

I.

THE GENEALOGY OF CREATION, newly Translated from the Unpointed Hebrew Text of the Book of Genesis, showing the General Scientific Accuracy of the Cosmogony of Moses and the Philosophy of Creation. 8vo. cloth, 14s.

II.

ON ECCENTRIC AND CENTRIC FORCE: A New Theory of Projection. With Engravings. 8vo. cloth, 10s.

III.

ON ORBITAL MOTION: The Outlines of a System of Physical Astronomy. With Diagrams. 8vo. cloth, 7s. 6d.

IV.

ASTRONOMICAL INVESTIGATIONS. The Cosmical Relations of the Revolution of the Lunar Apsides. Oceanic Tides. With Engravings. 8vo. cloth, 5s.

V.

THE ORACLES OF GOD: An Attempt at a Re-interpretation. Part I. The Revealed Cosmos. 8vo. cloth, 10s.

THE PRESCRIBER'S PHARMACOPŒIA; containing all the Medicines in the British Pharmacopœia, arranged in Classes according to their Action, with their Composition and Doses. By a Practising Physician. Fifth Edition. 32mo. cloth, 2s. 6d.; roan tuck (for the pocket), 3s. 6d.

DR. JOHN ROWLISON PRETTY.

AIDS DURING LABOUR, including the Administration of Chloroform, the Management of Placenta and Post-partum Hæmorrhage. Fcap. 8vo. cloth, 4s. 6d.

MR. P. C. PRICE, F.R.C.S.

AN ESSAY ON EXCISION OF THE KNEE-JOINT. With Coloured Plates. With Memoir of the Author and Notes by Henry Smith, F.R.C.S. Royal 8vo. cloth, 14s.

DR. PRIESTLEY.

LECTURES ON THE DEVELOPMENT OF THE GRAVID UTERUS. 8vo. cloth, 5s. 6d.

DR. RADCLIFFE, F.R.C.P.L.

LECTURES ON EPILEPSY, PAIN, PARALYSIS, AND CERTAIN OTHER DISORDERS OF THE NERVOUS SYSTEM, delivered at the Royal College of Physicians in London. Post 8vo. cloth, 7s. 6d.

MR. RAINEY.

ON THE MODE OF FORMATION OF SHELLS OF ANIMALS, OF BONE, AND OF SEVERAL OTHER STRUCTURES, by a Process of Molecular Coalescence, Demonstrable in certain Artificially-formed Products. Fcap. 8vo. cloth, 4s. 6d.

DR. F. H. RAMSBOTHAM.

THE PRINCIPLES AND PRACTICE OF OBSTETRIC MEDICINE AND SURGERY. Illustrated with One Hundred and Twenty Plates on Steel and Wood; forming one thick handsome volume. Fourth Edition. 8vo. cloth, 22s.

DR. RAMSBOTHAM.
PRACTICAL OBSERVATIONS ON MIDWIFERY, with a Selection of Cases. Second Edition. 8vo. cloth, 12s.

PROFESSOR REDWOOD, PH.D.
A SUPPLEMENT TO THE PHARMACOPŒIA: A concise but comprehensive Dispensatory, and Manual of Facts and Formulæ, for the use of Practitioners in Medicine and Pharmacy. Third Edition. 8vo. cloth, 22s.

DR. DU BOIS REYMOND.
ANIMAL ELECTRICITY; Edited by H. BENCE JONES, M.D., F.R.S. With Fifty Engravings on Wood. Foolscap 8vo. cloth, 6s.

DR. REYNOLDS, M.D.LOND.
I.
EPILEPSY: ITS SYMPTOMS, TREATMENT, AND RELATION TO OTHER CHRONIC CONVULSIVE DISEASES. 8vo. cloth, 10s.

II.
THE DIAGNOSIS OF DISEASES OF THE BRAIN, SPINAL CORD, AND THEIR APPENDAGES. 8vo. cloth, 8s.

DR. B. W. RICHARDSON.
I.
ON THE CAUSE OF THE COAGULATION OF THE BLOOD. Being the ASTLEY COOPER PRIZE ESSAY for 1856. With a Practical Appendix. 8vo. cloth, 16s.
II.
THE HYGIENIC TREATMENT OF PULMONARY CONSUMPTION. 8vo. cloth, 5s. 6d.

DR. RITCHIE, M.D.
ON OVARIAN PHYSIOLOGY AND PATHOLOGY. With Engravings. 8vo. cloth, 6s.

DR. WILLIAM ROBERTS, M.D., F.R.C.P.
AN ESSAY ON WASTING PALSY; being a Systematic Treatise on the Disease hitherto described as ATROPHIE MUSCULAIRE PROGRESSIVE. With Four Plates. 8vo. cloth, 5s.

DR. ROUTH.
INFANT FEEDING, AND ITS INFLUENCE ON LIFE; Or, the Causes and Prevention of Infant Mortality. Second Edition. Fcap. 8vo. cloth, 6s.

DR. W. H. ROBERTSON.
I.
THE NATURE AND TREATMENT OF GOUT. 8vo. cloth, 10s. 6d.
II.
A TREATISE ON DIET AND REGIMEN. Fourth Edition. 2 vols. 12s. post 8vo. cloth.

DR. ROWE.

NERVOUS DISEASES, LIVER AND STOMACH COMPLAINTS, LOW SPIRITS, INDIGESTION, GOUT, ASTHMA, AND DISORDERS PRODUCED BY TROPICAL CLIMATES. With Cases. Sixteenth Edition. Fcap. 8vo. 2s. 6d.

DR. ROYLE, F.R.S., AND DR. HEADLAND, M.D.

A MANUAL OF MATERIA MEDICA AND THERAPEUTICS. With numerous Engravings on Wood. Fourth Edition. Fcap. 8vo. cloth, 12s. 6d.

DR. RYAN, M.D.

INFANTICIDE: ITS LAW, PREVALENCE, PREVENTION, AND HISTORY. 8vo. cloth, 5s.

ST. BARTHOLOMEW'S HOSPITAL.

A DESCRIPTIVE CATALOGUE OF THE ANATOMICAL MUSEUM. Vol. I. (1846), Vol. II. (1851), Vol. III. (1862), 8vo. cloth, 5s. each.

ST. GEORGE'S HOSPITAL REPORTS. Vol. I. 8vo. 7s. 6d.

MR. T. P. SALT, BIRMINGHAM.

I.

ON DEFORMITIES AND DEBILITIES OF THE LOWER EXTREMITIES, AND THE MECHANICAL TREATMENT EMPLOYED IN THE PROMOTION OF THEIR CURE. With numerous Plates. 8vo. cloth, 15s.

II.

ON RUPTURE: ITS CAUSES, MANAGEMENT, AND CURE, and the various Mechanical Contrivances employed for its Relief. With Engravings. Post 8vo. cloth, 3s.

DR. SALTER, F.R.S.

ON ASTHMA: its Pathology, Causes, Consequences, and Treatment. 8vo. cloth, 10s.

DR. SANKEY, M.D.LOND.

LECTURES ON MENTAL DISEASES. 8vo. cloth, 8s.

DR. SANSOM, M.B.LOND.

I.

CHLOROFORM: ITS ACTION AND ADMINISTRATION. A Handbook. With Engravings. Crown 8vo. cloth, 5s.

II.

THE ARREST AND PREVENTION OF CHOLERA; being a Guide to the Antiseptic Treatment. Fcap. 8vo. cloth, 2s. 6d.

MR. SAVORY.

A COMPENDIUM OF DOMESTIC MEDICINE, AND COMPANION TO THE MEDICINE CHEST; intended as a Source of Easy Reference for Clergymen, and for Families residing at a Distance from Professional Assistance. Seventh Edition. 12mo. cloth, 5s.

DR. SCHACHT.

THE MICROSCOPE, AND ITS APPLICATION TO VEGETABLE ANATOMY AND PHYSIOLOGY. Edited by FREDERICK CURREY, M.A. Fcap. 8vo. cloth, 6s.

DR. SCORESBY-JACKSON, M.D., F.R.S.E.

MEDICAL CLIMATOLOGY; or, a Topographical and Meteorological Description of the Localities resorted to in Winter and Summer by Invalids of various classes both at Home and Abroad. With an Isothermal Chart. Post 8vo. cloth, 12s.

DR. SEMPLE.

ON COUGH : its Causes, Varieties, and Treatment. With some practical Remarks on the Use of the Stethoscope as an aid to Diagnosis. Post 8vo. cloth, 4s. 6d.

DR. SEYMOUR.

I.

ILLUSTRATIONS OF SOME OF THE PRINCIPAL DIS-EASES OF THE OVARIA: their Symptoms and Treatment; to which are prefixed Observations on the Structure and Functions of those parts in the Human Being and in Animals. On India paper. Folio, 16s.

II.

THE NATURE AND TREATMENT OF DROPSY; considered especially in reference to the Diseases of the Internal Organs of the Body, which most commonly produce it. 8vo. 5s.

DR. SHAPTER, M.D., F.R.C.P.

THE CLIMATE OF THE SOUTH OF DEVON, AND ITS INFLUENCE UPON HEALTH. Second Edition, with Maps. 8vo. cloth, 10s. 6d.

MR. SHAW, M.R.C.S.

THE MEDICAL REMEMBRANCER ; OR, BOOK OF EMER-GENCIES. Fourth Edition. Edited, with Additions, by JONATHAN HUTCHINSON, F.R.C.S. 32mo. cloth, 2s. 6d.

DR. SHEA, M.D., B.A.

A MANUAL OF ANIMAL PHYSIOLOGY. With an Appendix of Questions for the B.A. London and other Examinations. With Engravings. Foolscap 8vo. cloth, 5s. 6d.

DR. SHRIMPTON.

CHOLERA : ITS SEAT, NATURE, AND TREATMENT. With Engravings. 8vo. cloth, 4s. 6d.

DR. SIBSON, F.R.S.

MEDICAL ANATOMY. With coloured Plates. Imperial folio. Fasci-culi I. to VI. 5s. each.

DR. E. H. SIEVEKING.

ON EPILEPSY AND EPILEPTIFORM SEIZURES: their Causes, Pathology, and Treatment. Second Edition. Post 8vo. cloth, 10s. 6d.

DR. SIMMS.

A WINTER IN PARIS : being a few Experiences and Observations of French Medical and Sanitary Matters. Fcap. 8vo. cloth, 4s.

MR. SINCLAIR AND DR. JOHNSTON.

PRACTICAL MIDWIFERY : Comprising an Account of 13,748 Deli-veries, which occurred in the Dublin Lying-in Hospital, during a period of Seven Years. 8vo. cloth, 10s.

DR. SIORDET, M.B.LOND., M.R.C.P.

MENTONE IN ITS MEDICAL ASPECT. Foolscap 8vo. cloth, 2s. 6d.

MR. ALFRED SMEE, F.R.S.

GENERAL DEBILITY AND DEFECTIVE NUTRITION ; their Causes, Consequences, and Treatment. Second Edition. Fcap. 8vo. cloth, 3s. 6d.

DR. SMELLIE.

OBSTETRIC PLATES: being a Selection from the more Important and Practical Illustrations contained in the Original Work. With Anatomical and Practical Directions. 8vo. cloth, 5s.

MR. HENRY SMITH, F.R.C.S.
I.

ON STRICTURE OF THE URETHRA. 8vo. cloth, 7s. 6d.

II.

HÆMORRHOIDS AND PROLAPSUS OF THE RECTUM: Their Pathology and Treatment, with especial reference to the use of Nitric Acid. Third Edition. Fcap. 8vo. cloth, 3s.

III.

THE SURGERY OF THE RECTUM. Lettsomian Lectures. Fcap. 8vo. 2s. 6d.

DR. J. SMITH, M.D., F.R.C.S.EDIN.

HANDBOOK OF DENTAL ANATOMY AND SURGERY, FOR THE USE OF STUDENTS AND PRACTITIONERS. Fcap. 8vo. cloth, 3s. 6d.

DR. W. TYLER SMITH.

A MANUAL OF OBSTETRICS, THEORETICAL AND PRACTICAL. Illustrated with 186 Engravings. Fcap. 8vo. cloth, 12s. 6d.

DR. SNOW.

ON CHLOROFORM AND OTHER ANÆSTHETICS: THEIR ACTION AND ADMINISTRATION. Edited, with a Memoir of the Author, by Benjamin W. Richardson, M.D. 8vo. cloth, 10s. 6d.

MR. J. VOSE SOLOMON, F.R.C.S.

TENSION OF THE EYEBALL; GLAUCOMA: some Account of the Operations practised in the 19th Century. 8vo. cloth, 4s.

DR. STANHOPE TEMPLEMAN SPEER.

PATHOLOGICAL CHEMISTRY, IN ITS APPLICATION TO THE PRACTICE OF MEDICINE. Translated from the French of MM. Becquerel and Rodier. 8vo. cloth, reduced to 8s.

MR. PETER SQUIRE.
I.

A COMPANION TO THE BRITISH PHARMACOPÆIA. Third Edition. 8vo. cloth, 8s. 6d.

II.

THE PHARMACOPÆIAS OF THIRTEEN OF THE LONDON HOSPITALS, arranged in Groups for easy Reference and Comparison. 18mo. cloth, 3s. 6d.

DR. STEGGALL.
STUDENTS' BOOKS FOR EXAMINATION.

I.

A MEDICAL MANUAL FOR APOTHECARIES' HALL AND OTHER MEDICAL BOARDS. Twelfth Edition. 12mo. cloth, 10s.

II.

A MANUAL FOR THE COLLEGE OF SURGEONS; intended for the Use of Candidates for Examination and Practitioners. Second Edition. 12mo. cloth, 10s.

III.

GREGORY'S CONSPECTUS MEDICINÆ THEORETICÆ. The First Part, containing the Original Text, with an Ordo Verborum, and Literal Translation. 12mo. cloth, 10s.

IV.

THE FIRST FOUR BOOKS OF CELSUS; containing the Text, Ordo Verborum, and Translation. Second Edition. 12mo. cloth, 8s.

V.

FIRST LINES FOR CHEMISTS AND DRUGGISTS PREPARING FOR EXAMINATION AT THE PHARMACEUTICAL SOCIETY. Second Edition. 18mo. cloth, 3s. 6d.

MR. STOWE, M.R.C.S.

A TOXICOLOGICAL CHART, exhibiting at one view the Symptoms, Treatment, and Mode of Detecting the various Poisons, Mineral, Vegetable, and Animal. To which are added, concise Directions for the Treatment of Suspended Animation. Twelfth Edition, revised. On Sheet, 2s.; mounted on Roller, 5s.

MR. FRANCIS SUTTON, F.C.S.

A SYSTEMATIC HANDBOOK OF VOLUMETRIC ANALYSIS; or, the Quantitative Estimation of Chemical Substances by Measure. With Engravings. Post 8vo. cloth, 7s. 6d.

DR. SWAYNE.

OBSTETRIC APHORISMS FOR THE USE OF STUDENTS COMMENCING MIDWIFERY PRACTICE. With Engravings on Wood. Third Edition. Fcap. 8vo. cloth, 3s. 6d.

MR. TAMPLIN, F.R.C.S.E.

LATERAL CURVATURE OF THE SPINE: its Causes, Nature, and Treatment. 8vo. cloth, 4s.

DR. ALEXANDER TAYLOR, F.R.S.E.

THE CLIMATE OF PAU; with a Description of the Watering Places of the Pyrenees, and of the Virtues of their respective Mineral Sources in Disease. Third Edition. Post 8vo. cloth, 7s.

DR. ALFRED S. TAYLOR, F.R.S.

I.

THE PRINCIPLES AND PRACTICE OF MEDICAL JURISPRUDENCE. With 176 Wood Engravings. 8vo. cloth, 28s.

II.

A MANUAL OF MEDICAL JURISPRUDENCE. Eighth Edition. With Engravings. Fcap. 8vo. cloth, 12s. 6d.

III.

ON POISONS, in relation to MEDICAL JURISPRUDENCE AND MEDICINE. Second Edition. Fcap. 8vo. cloth, 12s. 6d.

MR. TEALE.

ON AMPUTATION BY A LONG AND A SHORT RECTAN-
GULAR FLAP. With Engravings on Wood. 8vo. cloth, 5s.

DR. THEOPHILUS THOMPSON, F.R.S.

CLINICAL LECTURES ON PULMONARY CONSUMPTION;
with additional Chapters by E. Symes Thompson, M.D. With Plates. 8vo. cloth, 7s. 6d.

DR. THOMAS.

THE MODERN PRACTICE OF PHYSIC; exhibiting the Symp-
toms, Causes, Morbid Appearances, and Treatment of the Diseases of all Climates.
Eleventh Edition. Revised by Algernon Frampton, M.D. 2 vols. 8vo. cloth, 28s.

MR. HENRY THOMPSON, F.R.C.S.

I.

STRICTURE OF THE URETHRA; its Pathology and Treatment.
The Jacksonian Prize Essay for 1852. With Plates. Second Edition. 8vo. cloth, 10s.

II.

THE DISEASES OF THE PROSTATE; their Pathology and Treat-
ment. Comprising a Dissertation "On the Healthy and Morbid Anatomy of the Prostate
Gland;" being the Jacksonian Prize Essay for 1860. With Plates. Second Edition.
8vo. cloth, 10s.

III.

PRACTICAL LITHOTOMY AND LITHOTRITY; or, An Inquiry
into the best Modes of removing Stone from the Bladder. With numerous Engravings,
8vo. cloth, 9s.

DR. THUDICHUM.

I.

A TREATISE ON THE PATHOLOGY OF THE URINE,
Including a complete Guide to its Analysis. With Plates, 8vo. cloth, 14s.

II.

A TREATISE ON GALL STONES: their Chemistry, Pathology,
and Treatment. With Coloured Plates. 8vo. cloth, 10s.

DR. TILT.

I.

ON UTERINE AND OVARIAN INFLAMMATION, AND ON
THE PHYSIOLOGY AND DISEASES OF MENSTRUATION. Third Edition.
8vo. cloth, 12s.

II.

A HANDBOOK OF UTERINE THERAPEUTICS, AND OF
MODERN PATHOLOGY OF DISEASES OF WOMEN. Second Edition.
Post 8vo. cloth, 6s.

III.

THE CHANGE OF LIFE IN HEALTH AND DISEASE: a
Practical Treatise on the Nervous and other Affections incidental to Women at the Decline
of Life. Second Edition. 8vo. cloth, 6s.

DR. GODWIN TIMMS.

CONSUMPTION: its True Nature and Successful Treatment. Crown
8vo. cloth, 10s.

DR. ROBERT B. TODD, F.R.S.

I.

CLINICAL LECTURES ON THE PRACTICE OF MEDICINE.
New Edition, in one Volume, Edited by DR. BEALE, *8vo. cloth,* 18s.

II.

ON CERTAIN DISEASES OF THE URINARY ORGANS, AND
ON DROPSIES. Fcap. 8vo. cloth, 6s.

MR. TOMES, F.R.S.

A MANUAL OF DENTAL SURGERY. With 208 Engravings on
Wood. Fcap. 8vo. cloth, 12s. 6d.

MR. JOSEPH TOYNBEE, F.R.S., F.R.C.S.

THE DISEASES OF THE EAR: THEIR NATURE, DIAG-
NOSIS, AND TREATMENT. Illustrated with numerous Engravings on Wood.
8vo. cloth, 15s.

DR. TURNBULL.

I.

AN INQUIRY INTO THE CURABILITY OF CONSUMPTION,
ITS PREVENTION, AND THE PROGRESS OF IMPROVEMENT IN THE
TREATMENT. Third Edition. 8vo. cloth, 6s.

II.

A PRACTICAL TREATISE ON DISORDERS OF THE STOMACH
with FERMENTATION; and on the Causes and Treatment of Indigestion, &c. 8vo.
cloth, 6s.

DR. TWEEDIE, F.R.S.

CONTINUED FEVERS: THEIR DISTINCTIVE CHARACTERS,
PATHOLOGY, AND TREATMENT. With Coloured Plates. 8vo. cloth, 12s.

VESTIGES OF THE NATURAL HISTORY OF CREATION.
Eleventh Edition. Illustrated with 106 Engravings on Wood. 8vo. cloth, 7s. 6d.

DR. UNDERWOOD.

TREATISE ON THE DISEASES OF CHILDREN. Tenth Edition,
with Additions and Corrections by HENRY DAVIES, M.D. 8vo. cloth, 15s.

DR. UNGER.

BOTANICAL LETTERS. Translated by Dr. B. PAUL. Numerous
Woodcuts. Post 8vo., 2s. 6d.

MR. WADE, F.R.C.S.

STRICTURE OF THE URETHRA, ITS COMPLICATIONS
AND EFFECTS; a Practical Treatise on the Nature and Treatment of those
Affections. Fourth Edition. 8vo. cloth, 7s. 6d.

DR. WALKER, M.B.LOND.
ON DIPHTHERIA AND DIPHTHERITIC DISEASES. Fcap. 8vo. cloth, 3s.

DR. WALLER.
ELEMENTS OF PRACTICAL MIDWIFERY; or, Companion to the Lying-in Room. Fourth Edition, with Plates. Fcap. cloth, 4s. 6d.

MR. HAYNES WALTON, F.R.C.S.
SURGICAL DISEASES OF THE EYE. With Engravings on Wood. Second Edition. 8vo. cloth, 14s.

DR. WARING, M.D., M.R.C.P.LOND.
I.
A MANUAL OF PRACTICAL THERAPEUTICS. Second Edition, Revised and Enlarged. Fcap. 8vo. cloth, 12s. 6d.
II.
THE TROPICAL RESIDENT AT HOME. Letters addressed to Europeans returning from India and the Colonies on Subjects connected with their Health and General Welfare. Crown 8vo. cloth, 5s.

DR. WATERS, M.R.C.P.
I.
THE ANATOMY OF THE HUMAN LUNG. The Prize Essay to which the Fothergillian Gold Medal was awarded by the Medical Society of London. Post 8vo. cloth, 6s. 6d.
II.
RESEARCHES ON THE NATURE, PATHOLOGY, AND TREATMENT OF EMPHYSEMA OF THE LUNGS, AND ITS RELATIONS WITH OTHER DISEASES OF THE CHEST. With Engravings. 8vo. cloth, 5s.

DR. ALLAN WEBB, F.R.C.S.L.
THE SURGEON'S READY RULES FOR OPERATIONS IN SURGERY. Royal 8vo. cloth, 10s. 6d.

DR. WEBER.
A CLINICAL HAND-BOOK OF AUSCULTATION AND PERCUSSION. Translated by JOHN COCKLE, M.D. 5s.

MR. SOELBERG WELLS, M.D., M.R.C.S.
ON LONG, SHORT, AND WEAK SIGHT, and their Treatment by the Scientific Use of Spectacles. Second Edition. With Plates. 8vo. cloth, 6s.

MR. T. SPENCER WELLS, F.R.C.S.

I.

DISEASES OF THE OVARIES: THEIR DIAGNOSIS AND
TREATMENT. Vol. I. 8vo. cloth, 9s.

II.

SCALE OF MEDICINES WITH WHICH MERCHANT VES-
SELS ARE TO BE FURNISHED, by command of the Privy Council for Trade;
With Observations on the Means of Preserving the Health of Seamen, &c. &c.
Seventh Thousand.   Fcap. 8vo. cloth, 3s. 6d.

DR. WEST.

LECTURES ON THE DISEASES OF WOMEN. Third Edition.
8vo. cloth, 16s.

DR. UVEDALE WEST.

ILLUSTRATIONS OF PUERPERAL DISEASES. Second Edi-
tion, enlarged.   Post 8vo. cloth, 5s.

MR. WHEELER.

HAND-BOOK OF ANATOMY FOR STUDENTS OF THE
FINE ARTS.   With Engravings on Wood. Fcap. 8vo., 2s. 6d.

DR. WHITEHEAD, F.R.C.S.

ON THE TRANSMISSION FROM PARENT TO OFFSPRING
OF SOME FORMS OF DISEASE, AND OF MORBID TAINTS AND
TENDENCIES. Second Edition. 8vo. cloth, 10s. 6d.

DR. WILLIAMS, F.R.S.

PRINCIPLES OF MEDICINE : An Elementary View of the Causes,
Nature, Treatment, Diagnosis, and Prognosis, of Disease. With brief Remarks on
Hygienics, or the Preservation of Health. The Third Edition. 8vo. cloth, 15s.

THE WIFE'S DOMAIN : the Young Couple—the Mother—the Nurse
—the Nursling.   Post 8vo. cloth, 3s. 6d.

DR. J. HUME WILLIAMS.

UNSOUNDNESS OF MIND, IN ITS MEDICAL AND LEGAL
CONSIDERATIONS. 8vo. cloth, 7s. 6d.

DR. WILLIAMSON, SURGEON-MAJOR, 64TH REGIMENT.

MILITARY SURGERY. With Plates. 8vo. cloth, 12s.

MR. ERASMUS WILSON, F.R.S.

I.

THE ANATOMIST'S VADE-MECUM: A SYSTEM OF HUMAN
ANATOMY. With numerous Illustrations on Wood. Eighth Edition. Foolscap 8vo.
cloth, 12s. 6d.

II.

DISEASES OF THE SKIN: A Practical and Theoretical Treatise on
the DIAGNOSIS, PATHOLOGY, and TREATMENT OF CUTANEOUS DIS-
EASES. Fifth Edition. 8vo. cloth, 16s.

THE SAME WORK; illustrated with finely executed Engravings on Steel, accurately
coloured. 8vo. cloth, 34s.

III.

HEALTHY SKIN: A Treatise on the Management of the Skin and Hair
in relation to Health. Seventh Edition. Foolscap 8vo. 2s. 6d.

IV.

PORTRAITS OF DISEASES OF THE SKIN. Folio. Fasciculi I.
to XII., completing the Work. 20s. each. The Entire Work, half morocco, £13.

V.

THE STUDENT'S BOOK OF CUTANEOUS MEDICINE AND
DISEASES OF THE SKIN. Post 8vo. cloth, 8s. 6d.

VI.

ON SYPHILIS, CONSTITUTIONAL AND HEREDITARY;
AND ON SYPHILITIC ERUPTIONS. With Four Coloured Plates. 8vo. cloth,
16s.

VII.

A THREE WEEKS' SCAMPER THROUGH THE SPAS OF
GERMANY AND BELGIUM, with an Appendix on the Nature and Uses of
Mineral Waters. Post 8vo. cloth, 6s. 6d.

VIII.

THE EASTERN OR TURKISH BATH: its History, Revival in
Britain, and Application to the Purposes of Health. Foolscap 8vo., 2s.

DR. G. C. WITTSTEIN.

PRACTICAL PHARMACEUTICAL CHEMISTRY: An Explanation
of Chemical and Pharmaceutical Processes, with the Methods of Testing the Purity of
the Preparations, deduced from Original Experiments. Translated from the Second
German Edition, by STEPHEN DARBY. 18mo. cloth, 6s.

DR. HENRY G. WRIGHT.

HEADACHES; their Causes and their Cure. Fourth Edition. Fcap. 8vo.
2s. 6d.

DR. YEARSLEY, M.D., M.R.C.S.

I.

DEAFNESS PRACTICALLY ILLUSTRATED; being an Exposition
as to the Causes and Treatment of Diseases of the Ear. Sixth Edition. 8vo. cloth, 6s.

II.

ON THROAT AILMENTS, MORE ESPECIALLY IN THE
ENLARGED TONSIL AND ELONGATED UVULA. Eighth Edition. 8vo.
cloth, 5s.

CHURCHILL'S SERIES OF MANUALS.

Fcap. 8vo. cloth, 12s. 6d. each.

"We here give Mr. Churchill public thanks for the positive benefit conferred on the Medical Profession, by the series of beautiful and cheap Manuals which bear his imprint."— *British and Foreign Medical Review.*

AGGREGATE SALE, 141,000 COPIES.

ANATOMY. With numerous Engravings. Eighth Edition. By ERASMUS WILSON, F.R.C.S., F.R.S.

BOTANY. With numerous Engravings. By ROBERT BENTLEY, F.L.S., Professor of Botany, King's College, and to the Pharmaceutical Society.

CHEMISTRY. With numerous Engravings. Ninth Edition. By GEORGE FOWNES, F.R.S., H. BENCE JONES, M.D., F.R.S., and A. W. HOFMANN, F.R.S.

DENTAL SURGERY. With numerous Engravings. By JOHN TOMES, F.R.S.

MATERIA MEDICA. With numerous Engravings. Fourth Edition. By J. FORBES ROYLE, M.D., F.R.S., and FREDERICK W. HEADLAND, M.D., F.L.S.

MEDICAL JURISPRUDENCE. With numerous Engravings. Eighth Edition. By ALFRED SWAINE TAYLOR, M.D., F.R.S.

PRACTICE OF MEDICINE. Second Edition. By G. HILARO BARLOW, M.D., M.A.

The MICROSCOPE and its REVELATIONS. With numerous Plates and Engravings. Third Edition. By W. B. CARPENTER, M.D., F.R.S.

NATURAL PHILOSOPHY. With numerous Engravings. Fifth Edition. By GOLDING BIRD, M.D., M.A., F.R.S., and CHARLES BROOKE, M.B., M.A., F.R.S.

OBSTETRICS. With numerous Engravings. By W. TYLER SMITH, M.D., F.R.C.P.

OPHTHALMIC MEDICINE and SURGERY. With coloured Plates and Engravings on Wood. Third Edition. By T. WHARTON JONES, F.R.C.S., F.R.S.

PATHOLOGICAL ANATOMY. With numerous Engravings. By C. HANDFIELD JONES, M.B., F.R.S., and E. H. SIEVEKING, M.D., F.R.C.P.

PHYSIOLOGY. With numerous Engravings. Fourth Edition. By WILLIAM B. CARPENTER, M.D., F.R.S.

POISONS. Second Edition. By ALFRED SWAINE TAYLOR, M.D., F.R.S.

PRACTICAL ANATOMY. With numerous Engravings. (10s. 6d.) By CHRISTOPHER HEATH, F.R.C.S.

PRACTICAL SURGERY. With numerous Engravings. Fourth Edition. By Sir WILLIAM FERGUSSON, Bart., F.R.C.S., F.R.S.

THERAPEUTICS. Second Edition. By E. J. Waring, M.D., M.R.C.P.

Printed by W. BLANCHARD & SONS, 62, Millbank Street, Westminster.

www.ingramcontent.com/pod-product-compliance
Lightning Source LLC
Chambersburg PA
CBHW021509210326
41599CB00012B/1194